Simulation of
Local Area Networks

Simulation of

Local Area Networks

Matthew N. O. Sadiku, Ph.D.
Associate Professor
Department of Electrical Engineeing
Temple University
Philadelphia, Pennsylvania

Mohammad Ilyas, Ph.D.
Professor
Department of Computer Science and Engineering
Florida Atlantic University
Boca Raton, Florida

CRC Press
Boca Raton Ann Arbor London Tokyo

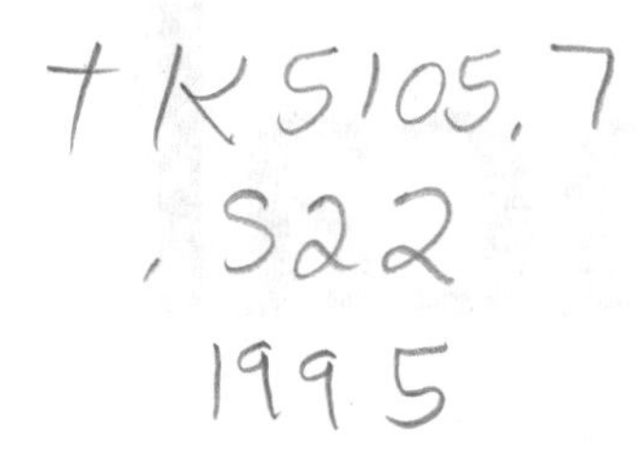

Library of Congress Cataloging-in-Publication Data

Sadiku, Matthew N. O.
Simulation of local area networks / Matthew N. O. Sadiku, Mohammad Ilyas
p. cm.
Includes bibliographical references and index.
ISBN 0-8493-2473-4
1. Local area networks (Computer networks)--Mathematical models
I. Ilyas, Mohammad, 1953- II. Title.
TK5105.7.S22 1994
004.6'8'01135133—dc20
DNLM/DLC
for Library of Congress 94-23413
CIP

International Standard Book Number 0-8493-2473-4
Library of Congress Card Number 94-23413
Printed in the United States of America 1 2 3 4 5 6 7 8 9 0
Printed on acid-free paper

Dedicated to our families:

Chris, Ann, and Joyce

Parveen, Safia, Omar, and Zakia

PREFACE

One of the growing areas in the communication industry is the internetworking of the increased proliferation of computers, particularly via local area networks (LANs). A LAN is a data communication system, usually owned by a single organization, that enables similar or dissimilar digital devices to talk to each other over a common transmission medium.

Establishing the performance characteristics of a LAN before putting it into use is of paramount importance. It gives the designer the freedom and flexibility to adjust various parameters of the network at the planning stage. This way the designer eliminates the risk of unforeseen bottlenecks, underuse or overuse of resources, and failure to meet targeted system requirements.

The common approaches to performance evaluation apply analytical models, simulation models, or hybrid models. Simulation models allow systems analysts to evaluate the performance of existing or proposed systems under different conditions that lie beyond what analytic models can handle. Unlike analytic models, simulation models can provide estimates of virtually any network performance measure.

So far there are two categories of texts on simulation. Texts in the first category cover the general principles of simulation and apply them to a specific system for illustration purposes; they are not specific enough to help a beginner apply those principles in developing a simulation model for a LAN. Texts in the second category instruct readers on how to apply special-purpose languages (such as GPSS, GASP, SIMSCRIPT, SIMULA, SLAM, and RESQ) in constructing a simulation model for computer systems including LANs. Besides the fact that these languages are still evolving and are limited, the reader must first learn the specific simulation language before a simulation model can be developed.

This text has two major advantages over these existing texts. First, it uses C, a well developed general-purpose language that is familiar to most analysts. This avoids the need for learning a new simulation language or package. Second, the text specifically applies the simulation principles to local area networks. In addition, the text is student oriented and is suitable for classroom use or self-learning.

The text is intended for LAN designers who want to analyze the performance of their designs using simulation. It may be used for a one-semester course on simulation of LANs. The main requirements for students taking such a course are introductory LAN course and a knowledge of a high-level language, preferably C. Although familiarity with probability theory and statistics is useful, it is not required.

The book consists of eight chapters. Each chapter has a list of references to the literature, and there is bibliography at the end of the book. Chapter 1 provides a brief review of local area networks, and Chapter 2 gives the analytical models of popular LANs—token-passing bus and ring networks, CSMA/CD, and star network. Chapter 3 covers the general principles of simulation, and Chapter 4 deals with fundamental concepts in probability and statistic relating to simulation modeling. Materials in Chapters 3 and 4 are specifically applied in developing simulation models on token-passing LANs, CSMA/CD LANs, and Star LANs in Chapters 5, 6, and 7 respectively. The computer codes in Chapters 5 to 7 are divided into segments and a detailed explanation of each segment is presented to give a thorough understanding of the simulation models. The entire codes are put together in the appendices. It is hoped that the ideas gained in learning how to simulate these common LANs can be applied to other communication systems.

The authors are indebted to various students and colleagues who have contributed to this book. We are particularly indebted to George Paramanis and Sharuhk Murad for working on some of the simulation models as special graduate projects. Special thanks are due to Robert Stern of CRC Press for providing expert editorial guidance on the manuscript. Finally, we owe much to our families for their patience and support while preparing the material. To them this book is dedicated.

Table of Contents

6 SIMULATION OF CSMA/CD LANS 107

7 SIMULATION OF STAR LANs 135

8 SIMULATION LANGUAGES 151

Chapter 1

Local Area Networks

Successful people make decisions quickly as soon as all the facts are available and change them very slowly if ever. Unsuccessful people make decisions very slowly, and change them often and quickly.

—Napoleon Hill

In order to fully participate in the information age, one must be able to communicate with others in a multitude of ways. Everyone is familiar with telephone communications and the use of television as a medium for transmitting information. However, the greatest interest today is centered on computer generated data, and its transmission has become the most rapidly developing facet of the communication industry. The overall communication problem may be viewed as involving three types of networks:

- Local area networks (LANs) providing communications over a relatively small area
- Metropolitan area networks (MANs) operating over a few hundred kilometers
- Wide area networks (WANs) providing communications over several kilometers, across the nation, or around the globe

WANs, such as the ARPANET in the US, have existed for several years, but the need for LANs has been identified much more recently. Although the two types of communication networks employ identical principles, their characteristics are quite different. In a WAN, which may span continents, the transmission media are relatively expensive because of the large extent of the networks. Transmission rates in a WAN may range from 2,400 to about 50,000 bits per second, whereas in a LAN they are much higher, typically from 1 to 10 million bits per second.

In a WAN, the data arrival rate is low enough to permit processors to ensure error-free transmission and message integrity. This is not the case with a LAN because of its much higher data rate. Typically WANs use the existing telephone network for communications (or more recently the national packet data network), whereas LANs use privately installed coaxial cable or twisted-pair wires.

In this and the next chapter, we briefly review the fundamentals of local area networks necessary for the rest of the book. The material in the chapter is also discussed in many journal papers and textbooks which are given in the reference list at the end of the chapter.

1.1 Definition of a LAN

A LAN is a data communication system, usually owned by a single organization, that allows similar or dissimilar digital devices to talk to each other over a common transmission medium. According to the IEEE,

> *A local area network is distinguished from other types of data networks in that communication is usually confined to a moderate geographic area such as a single office building, a warehouse, or a campus, and can depend on a physical communications channel of moderate-to-high data rate which has a consistently low error rate.*

Thus we may regard a LAN as a resource-sharing data communication network with the following characteristics [1, 2]:

- Short distances (0.1 to 10 km)
- High speed (1 to 16 Mbps)
- Low cost (in the region of $3,000)
- Low error rate (10^{-8} to 10^{-11})
- Ease of access
- High reliability/integrity.

The network may connect data devices such as computers, terminals, mass storage devices, and printers/plotters. Through the network these devices can interchange data information such as file transfer, electronic mail, and word processing.

1.2 Evolution of LANs

LANs, as data communication networks, resulted from the marriage of two different technologies: telecommunications and computers. Data communication takes advantage of CATV technology to produce better performance at lower costs. Recent developments of large scale and very large scale (LSI and VLSI) integrated circuits have rapidly reduced the cost of computation and memory hardware. This has resulted in widely available low-cost personal computers, intelligent terminals, workstations, and minicomputers. However, other expensive resources such as high-quality printers, graphic plotters, and disk storage are best shared in a geographically limited area using a LAN.

Research in LANs began in the early 1970s, spurred by increasing requirements for resource sharing in multiple processor environments. In many cases these requirements first appeared in university campuses or research laboratories [3]. Ethernet, the first bus contention technology, originated at Xerox Corporation's Palo Alto Research Center, in the mid-1970s [4]. Called *Ethernet* after the concept in classical physics of wave transmission through an ether, the design borrowed many of the techniques and characteristics of the Aloha network, a packet radio network developed at the University of Hawaii. Since the introduction of Ethernet, networks using a number of topologies and protocols have been developed and reported. Typical examples are the token-ring topology developed in the US, mainly at MIT, but now the subject of IBM development work in Europe, and the Cambridge ring, which was produced at Cambridge University Computer Laboratory in the UK. The 1980s have been a decade of rapid maturation for LANs.

LANs represent a comparatively new field of activity and continue to hold the public interest. This is mirrored by the numerous courses being offered in the subject, by the many conference sessions devoted to LANs, by the research and development work on LANs being pursued both at the universities and in industries, and by the increasing amount of literature devoted to LANs.

All this interest is generated by the LAN's promise as a means of interconnecting various computers or computer-related devices into systems that are more useful than their individual parts. The goal of LANs is to provide a large number of devices with inexpensive yet high-speed local communications.

Communication between computers is becoming increasingly important as data processing becomes a commodity. Local area networking is a very rapidly growing field. Continued efforts are being made for further technological developments and innovations in the organization of these networks for maximal operational efficiency.

Today's LANs are on the edge of broadband speeds, and new LAN proposals call for higher speeds—e.g., a proposed 16 Mbps token ring LAN and a 100 Mbps fiber distributed data interface (FDDI). Higher speeds are needed to match the increasing speed of the PCs and to support diskless workstations and computer-aided design/manufacturing (CAD/CAM) terminals that rely on LANs for interconnection of file servers. Also, the success of LANs has led to several attempts to extend high-speed data networking beyond the local premises, across metropolitan and wide area environments.

1.3 LAN Technology

The types of technologies used to implement LANs are as diverse as the 200 or so LAN vendors. Both vendors and users are compelled to make a choice. This choice is usually based on several criteria such as [5–10]:

- Network topology and architecture
- Access control
- Transmission medium
- Transmission technique
- Adherence to standards
- Performance in terms of channel utilization, delay, power, and effective transmission ratio

The first four criteria are the major technological issues and are of great concern when discussing LANs. These issues are sometimes interrelated. For example, some access methods are only suitable for some topologies or with certain transmission techniques.

1.3.1 Network Topology and Access Control

The topology of a network is the way in which the nodes (or stations) are interconnected. In spite of the proliferation of LAN products, the vast majority of LANs conform to one of three topologies and one of a handful of medium-access control protocols summarized in Table 1.1. The basic forms of the topologies are shown in Figure 1.1.

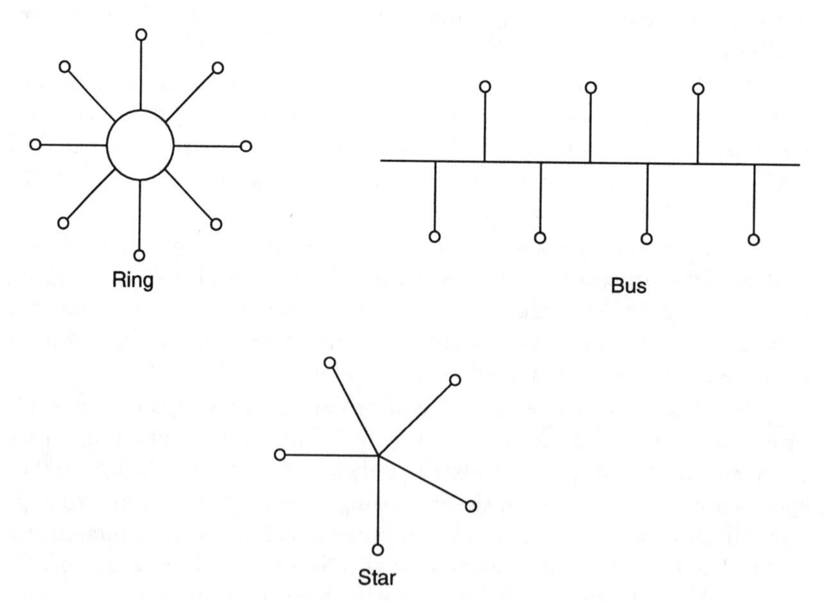

Figure 1.1 Local network topologies.

In the ring topology, all nodes are connected together in a closed loop. Information passes from node to node on the loop and is regenerated (by repeaters) at each node (called an *active* interface). A bus topology uses a single, open-ended transmission medium. Each node taps into the medium in a way that does not disturb the signal on the bus (thus it is called a *passive* interface). Star topology consists of a central controlling node with star-like connections to various other nodes. Although the ring topology is popular in Europe, the bus is the most common topology in the US.

Table 1.1 LAN Topology and Medium-Access Protocol.

1. Ring topology
   - Controlled Access
     - Token
     - Slotted
     - Buffer/Register Insertion
2. Bus topology
   - Controlled Access
     - Token
     - Multilevel Multiple Access (MLMA)
   - Random Access
     - Carrier Sense Multiple Access (CSMA) (1-persistent, *p*-persistent, nonpersistent)
     - Carrier Sense Multiple Access with Collision Detection (CSMA/CD)
     - CSMA/CD with Dynamic Priorities (CSMA/CD-CP)
     - CSMA/CD with Deterministic Contention Resolution (CSMA/CD-DCR)
3. Star topology
4. Hybrid topology
   - Ring–star
   - Ring–bus
   - Bus–star
   - CSMA/CD–token ring
   - CSMA/CD–token bus

Both ring and bus topologies, lacking any central node, must use some distributed mechanism to determine which node may use the transmission medium at any given moment. Various flow control and access strategies have been proposed or developed for inserting and removing messages from ring and bus LANs. The most popular ones are the Carrier Sense Multiple Access with Collision Detection (CSMA/CD) for broadcast buses and token passing for rings and buses.

In CSMA/CD, each bus interface unit (BIU), before attempting to transmit data onto the channel, first listens or senses if the channel is idle. An active BIU transmits its data only if the channel is sensed idle. If the channel is sensed busy, the BIU defers its transmission until the bus becomes clear. In this contention-type access scheme, collision occurs when two or more nodes attempt to transmit at the same time. During the collision, the two or more messages become garbled and lost.

In the token-passing protocol, an empty or idle token (some unique bit pattern or signal) is passed around the ring or bus. Any node may remove the token, insert a message, and append the token. When a node has data to transmit, it grabs the token, changes the token to a busy state (another bit pattern) and appends its packet to the busy token. At the end of the transmission, the node issues another idle token. A node has channel access right only when it gets the idle token. Figure 1.2 shows the packet format for token bus and token ring topologies.

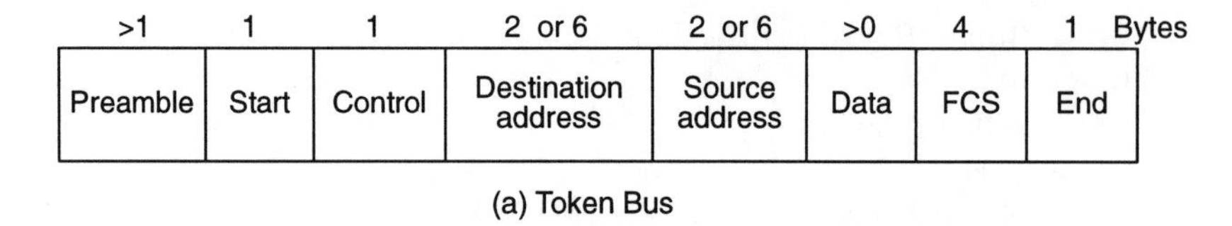

(a) Token Bus

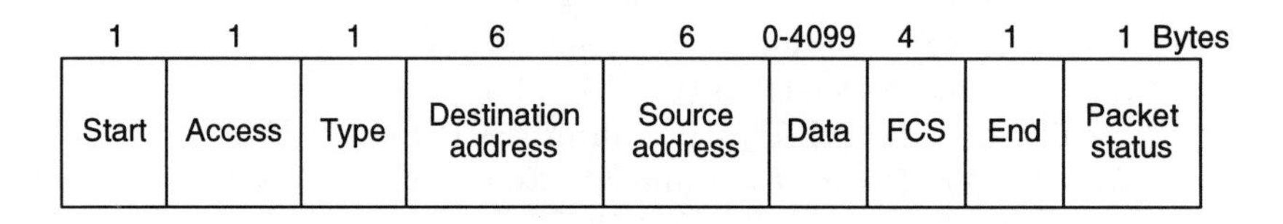

(b) Token Ring

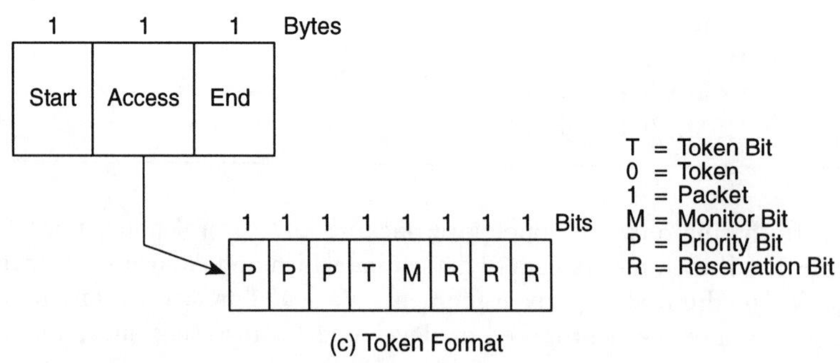

(c) Token Format

Figure 1.2 Packet format for token bus and token ring.

A number of attempts have been proposed to develop hybrid topologies and access techniques that combine random and time-assignment access.

1.3.2 Transmission Media

The transmission medium is the physical path connecting the transmitter to the receiver. Any physical medium that is capable of carrying information in an electromagnetic form is potentialy suitable for use on a LAN. In practice, the media used are twisted-pair cable, coaxial cable, and optical fiber.

Twisted-pair cable is generally used for analog signals but has been successfully used for digital transmissions. It is limited in speed to a few megabits and is often susceptible to noise. Ring, bus, and star networks can all use twisted-pair cable as a medium.

Coaxial cable consists of a central conductor and a conducting shield. It provides a substantial performance improvement over twisted-pair: it has higher capacity, can support a larger number of devices, and can span greater distances.

Optical fiber transmits light or infrared rays instead of electrical signals. It demonstrates higher capacity than coaxial cable and is not susceptible to noise or electrical fluctuations. Although there are still some technical difficulties with optical fiber, it may well be the medium choice of future networks.

1.3.3 Transmission Techniques

There are two types of transmission techniques: baseband signaling and broadband signaling. In spite of the hot debate and controversy about the merits of one technique over the other, it appears that the two techniques will coexist for some time, filling different needs.

Baseband signaling literally means that the signal is not modulated at all. It is totally digital. The entire frequency spectrum is used to form the signal, which is transmitted bidirectionally on broadcast systems such as buses. Baseband networks are limited in distance due to signal attenuation.

Broadband signaling is a technique by which information is frequency modulated onto analog carrier waves. This allows voice, video, and data to be carried simultaneously. Although it is more expensive than baseband because of the need for modems at each node, it provides larger capacity.

1.4 Standardization of LANs

The incompatibility of LAN products has left the market small and undecided. One way to increase the market size is to develop standards that can be used by the numerous LAN product manufacturers. The most obvious advantage of standards is that they facilitate the interchange of data between diverse devices connected to LAN. The driving force behind the standardization efforts is the desire by LAN users and vendors to have "open systems" in which any standard computer device would be able to interoperate with others. Attempts to standardize LAN topologies, protocols, and modulation techniques have been made by organizations such as those shown in Table 1.2.

Table 1.2 LAN Standardization Activities

Organization	Standards
ISO	TC 97/SC6 CSMA/CD, token bus, token ring, slotted ring
CCITT	SG 18 Connection between ISDN and LAN
IEEE	802 Logical link control, CSMA/CD, token bus, token ring, metropolitan area network
ECMA	TC 24 CSMA/CD, token bus, token ring

The International Standards Organization (ISO) is perhaps the most prominent of these, and it is responsible for the seven-layer model of network architecture, initially developed for WANs, called the reference model for open systems interconnection (OSI). The International Telegraph and Telephone Consultative Committee (CCITT), based in Geneva, is a part of the International Telecommunications Union (ITU) and is heavily involved in all aspects of data transmission. It has produced standards in Europe but not in the US. In the US the American National Standard Institute (ANSI), the National Bureau of Standards (NBS), and the IEEE are perhaps the most important bodies. The IEEE 802 committee has developed LAN protocol standards for the lower two layers of the ISO's OSI reference model, namely the physical and link control layers. The physical standard defines what manufacturers must provide in terms of access to their hardware for a user or systems integrator. More recently, the European Computer Manufacturers Association (ECMA) has followed along the same lines. Although network standards are being developed by the various organizations, standardization is still up to the manufacturers. The ISO protocols have the advantage of international backing, and most manufacturers have made the commitment to implement them eventually.

1.5 LAN Architectures

A network architecture is a specification of the set of functions required for a user at a location to interact with another user at another location. These interconnect functions include determination of the start and end of a message, recognition of a message address, management of a communication link, detection and recovery of transmission errors, and reliable and regulated delivery of data. A general network architecture thus describes the interfaces, algorithms, and protocols by which processes at different locations and/or at heterogeneous machines could communicate. Although the architectures developed by different vendors are functionally equivalent, they do not provide for easy interconnection of systems of different make.

1.5.1 The OSI Model

As mentioned in the previous section, the need for standardization and compatibility at all levels has compelled the International Standards Organization (ISO) to establish a general seven-layer hierarchical model for universal intercomputer communication. This arhitecture, known as the Open Systems Interconnection model (see Figure 1.3), defines seven layers of communication protocols, with specific functions isolated at each level. The OSI model is a reference model for the exchange of information among systems that are *open* to one another for this purpose by virtue of their mutual use of the applicable standards. The seven hierarchical layers are hardware and software functional groupings with specific well-defined tasks. The OSI model states the purpose of each layer and describes the services provided by each within its layer and to the adjacent higher and lower layers. Details of the implementation of each layer of OSI model depend on the specifics of the application and the characteristics of the communication channel employed.

1.5.2 The Seven OSI Layers

The *application layer*, level 7, is the one the user sees. It provides services directly comprehensible to application programs: login, password checks, network transparency for distribution of resources, file and document transfer, and industry-specific protocols.

The *presentation layer*, level 6, is concerned with interpreting the data. It restructures data to or from the standardized format used within a network, text compression, code conversion, file format conversion, and encryption.

The *session layer*, level 5, manages address translation and access security. It negotiates to establish a connection with another node on the network and then to manage the dialogue. This means controlling the starting, stopping, and synchronization of the conversion.

The *transport layer*, level 4, performs error control, sequence checking, handling of duplicate packets, flow control, and multiplexing. Here it is determined whether the channel is to be point-to-point (virtual) with ordered messages, isolated messages with no order, or broadcast messages. It is the last of the layers concerned with communications between peer entities in the systems. The transport layer and those above are referred to as the upper layers of the model, and they are independent of the underlying network. The lower layers are concerned with data transfer across the network.

The *network layer*, level 3, provides a connection path between systems, including the case where intermediate nodes are involved. It deals with message packetization, message routing for data transfer between nonadjacent nodes or stations, congestion control, and accounting.

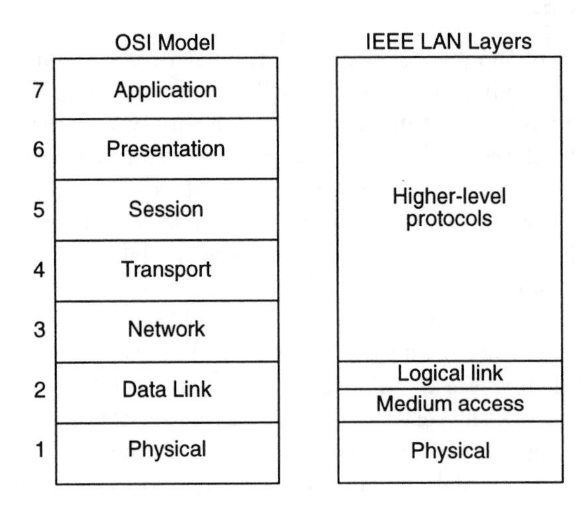

Figure 1.3 Relationship between the OSI model and IEEE LAN layers.

The *data link layer*, level 2, establishes the transmission protocol, how information will be transmitted, acknowledgment of messages, token possession, error detection, and sequencing. It prepares the packets passed down from the network layer for transmission on the network. It takes a raw transmission and transforms it into a line free from error. Here headers and framing information are added or removed. With these go the timing signals, check-sum, and station addresses, as well as the control system for access.

The *physical layer*, level 1, is that part that actually touches the transmission medium or cable; the line is the point within the node or device where the data is received and transmitted. It ensures that ones arrive as ones and zeros as zeros. It encodes and physically transfers messages (raw bits in a stream) between adjacent stations. It handles voltages, frequencies, direction, pin numbers, modulation techniques, signaling schemes, ground loop prevention, and collision detection in the CSMA/CD access method.

1.5.3 The IEEE Model for LANs

The IEEE has formulated standards for the physical and logical link layers for three types of LANs, namely, token buses, token rings, and

CSMA/CD protocols. Figure 1.3 illustrates the correspondence between the three layers of the OSI and the IEEE 802 reference models. The physical layer specifies means for transmitting and receiving bits across various types of media. The media-access-control layer performs the functions needed to control access to the physical medium. The logical-link-control layer is the common interface to the higher software layers.

1.6 Performance Evaluation

There has been much research into the performance of the various medium access protocols [5, 6, 11, 12]. LANs being complex systems, modeling of them must be done at various levels of detail. In this section, we present simple performance models of LANs. The performance is measured in terms of channel utilization, delay, power, and effective transmission ratio. More rigorous performance models of LANs are covered in the next chapter.

1.6.1 Channel Utilization

A performance yardstick is the maximum throughput achievable for a given channel capacity. For example, how many megabits per second (Mbps) of data can actually be transmitted for a channel capacity of 10 Mbps? It is certain that a fraction of the channel capacity is used up in form of overhead—acknowledgments, retransmission, token delay, etc.

Channel capacity is the maximum possible data rate, that is, the signaling rate on the physical channel. It is also known as the data rate or transmission rate and will be denoted by R in bits per second. Throughput S is the amount of "user data" that is carried by the LAN. Channel utilization U is the ratio of throughput to channel capacity—i.e.

$$U = \frac{S}{R} \tag{1.1}$$

It is independent of the medium access control. It is obvious that $U = 1$ in an ideal situation.

In analyzing LAN performance, the two most useful parameters are the channel capacity or data rate R of the medium and the average maximum signal propagation delay P. Their product (RP, in bits) is the number of bits that can exist in the channel between two nodes separated by the maximum distance determined by the propagation delay. We define the ratio

$$\alpha = \frac{\text{Length of data path}}{\text{Length of packet}} = \frac{RP}{P_L} \tag{1.2}$$

where P_L is the packet length in bits. The quantity α is a normalized nondimensional measure used in determining the upper bound on utilization; its reciprocal is called the effective transmission ratio. Realizing that P_L/R is the time needed to transmit a packet,

$$\alpha = \frac{P}{P_L/R} = \frac{\text{Propagation time}}{\text{Transmission time}} \tag{1.3}$$

If the throughput S is defined as the actual number of bits transmitted per second, then

$$S = \frac{\text{Length of packet}}{\text{Transmission time} + \text{Propagation time}}$$

or

$$S = \frac{P_L}{(P_L/R) + P} \tag{1.4}$$

Substituting (1.4) into (1.1) gives

$$U = \frac{P_L}{P_L + RP}$$

or

$$U = \frac{1}{1+\alpha} \tag{1.5}$$

Thus utilization is inversely related to α. Typical values of α range from 0.01 to 0.1 according to Stallings [13]. The ideal case occurs when there is no overhead and $\alpha = 0$. The ratio α can now be used to define channel utilization bounds for a medium access protocol. For token ring or bus,

$$U = \frac{T_1}{T_1 + T_2} = \frac{1}{1 + (T_2/T_1)} \tag{1.6}$$

where T_1 is the packet transmission period and T_2 is a token transmission period. It can be shown that [6, 13]

$$U = \begin{cases} \dfrac{1}{1+(\alpha/N)} & \alpha < 1 \\ \dfrac{1}{\alpha+(\alpha/N)} & \alpha > 1 \end{cases} \tag{1.7}$$

where N is the number of active nodes or stations. In a token bus, the optimal case occurs when the logical ordering of nodes (the sequence of nodes through which the token is passed, as shown in Figure 2.4) is the same as the physical order. In this case, Eq. (1.7) applies. In the worst case, the logical ordering of nodes forces the propagation delay between nodes to approach the end-to-end delay. For this case, $T_2 = \alpha$ and

$$U = \begin{cases} \dfrac{1}{1+\alpha} & \alpha < 1 \\ \dfrac{1}{2\alpha} & \alpha > 1 \end{cases} \tag{1.8}$$

1.6.2 Delay, Power, and Effective Transmission Ratio

Packet delay is the period of time between the moment at which a node becomes active (i.e., when it has data to transmit) and the moment at which the packet is successfully transmitted. Throughput delay describes the trade-off between throughput and packet delay. Delay D is the sum of the service time S plus the time W spent waiting to transmit all messages queued ahead of it and the actual propagation delay T_p. Thus

$$D = W + S + T_p \tag{1.9}$$

A performance measure combining throughput and delay into single function is the notion of power introduced in Gail and Kleinrock [14]. Power has recently evolved as a potentially useful measure of computer network performance in that it suggests an appropriate operating point for single networks. It is simply defined as

$$p = \frac{\lambda}{D} \tag{1.10}$$

where λ is the total arrival rate of messages to the network. One might consider that an appropriate operating point for a network is to choose that value of λ which maximizes power. From (1.10),

$$D\frac{dp}{d\lambda} + p\frac{dD}{d\lambda} = 1$$

Hence $\frac{dp}{d\lambda} = 0$ when

$$\lambda\frac{dD}{d\lambda} = D \tag{1.11}$$

that is, the optimum power point defines the "knee" of the $D(\lambda)$ curve. Thus, the optimum power point occurs at that value of λ where a straight line through the origin in the $D - \lambda$ plane is tangent to the D curve.

Another useful performance criterion is the effective transmission ratio, defined as

$$\text{Eff} = \frac{P_L}{R\ D} \tag{1.12}$$

where P_L is mean packet (or message) length in bits. Eff is a normalized nondimensional performance measure.

Example 1.1

Consider a bus LAN with 10 stations, an average internodal distance of 200 m, a transmission rate of 5 Mbps, and a packet size of 1,000 bits.

If the propogation speed is 2.0×10^8 m/s, calculate the throughput and channel utilization.

Solution

The distance ℓ between the two stations at the extreme ends is

$$\ell = 200(N-1) = 200 \times 9 = 1{,}800 \text{ m}$$

Hence the propagation time is

$$P = \frac{\ell}{u} = \frac{1{,}800}{2 \times 10^8} = 9 \times 10^{-6} \text{ s}$$

The packet transmission time is

$$T = \frac{P_L}{R} = \frac{1{,}000}{5 \times 10^6} = 0.5 \times 10^{-3} \text{ s}$$

Channel utilization is

$$U = \frac{P}{T} = \frac{9 \times 10^{-6}}{0.5 \times 10^{-3}} = 0.018$$

Thus, the throughput is obtained as

$$S = UR = 0.018 \times 5 \times 10^6 = 90 \text{ kbps}$$

1.7 Summary

The need to communicate—to send messages or data, to share expensive resources, to access computing facilities—has contributed to the development and spread of local area computer networks. A local area computer network is a system of computer-based stations interconnected by communication links. It provides a point-to-point communication among these stations located within a moderately sized geographical area. Communication takes place at data rates of 0.1 Mbps to 16 Mbps. Transmission can be baseband or broadband.

Commonly used transmission media employed with LANs include twisted-pairs wire, coaxial cable, and optical fiber. Fiber optics will be the transmission medium of the future because of its high bandwidth capability and reliability.

The three common topologies used for LANs are bus, ring, and star. Access to the transmission medium can be controlled or random. Token ring and token bus topologies use token passing protocol, whereas carrier sense multiple access with collision detection (CSMA/CD) uses random control on a bus- or tree-structured network.

Network architectures use a layered approach and define the interfaces between layers in a given network node and within the same layer in two different nodes. OSI provides a generalized model of system interconnection, and IEEE Project 802 has developed a set of standards for LANs.

Remarkable progress has been made in the field of computer communication networks with the acceptance of protocols and standards. The capabilities of LANs and the benefits they offer are clear and their future is bright. Local area networks will assume a monumental role in our future lives and will have a lasting impact on the way we conduct business transactions.

References

[1] C. D. Tsao, "A Local Area Network Architecture Overview," *IEEE Comm. Mag.*, vol. 22, no. 8, Aug. 1984, pp. 7–11.

[2] W. Stallings, *Local and Metropolitan Area Networks.* New York: Macmillan, 4th ed., 1993.

[3] J. M. Kryskow and C. K. Miller, "Local Area Networks Overview (2 parts)," *Computer Design*, Feb. 1981, pp. 22–25, Mar. 1981, pp. 12–20.

[4] R. M. Metcalfe and D. R. Boggs, "Ethernet: Distributed Packet Switching For Local Computer Networks," *Comm. of ACM*, vol. 19, no. 7, July 1976, pp. 395–404.

[5] J. L. Hammond and P. J. P. O'Reilly, *Performance Analysis of Local Computer Networks.* Reading, MA: Addison-Wesley, 1986.

[6] W. J. Neilson and U. M. Maydell, "A Survey of Current LAN Technology and Performance," *Infor*, vol. 23, no. 3, Aug. 1985, pp. 215–247.

[7] R. I. Wittlin and D. V. Ratner, "Choosing The Best Local Area Network for any Application," *Computer Design*, vol. 24, no. 2, Feb. 1985, pp. 143–149.

[8] L. Reiss, *Introduction to Local Area Networks with Microcomputer and Experiments.* Englewood Cliffs, NJ: Prentice-Hall, 1987, pp. 15–34.

[9] K. C. E. Gee, *Introduction to Local Area Computer Networks.* New York: Wiley, 1983, chaps. 3 and 4, pp. 11–52.

[10] W. Stallings, *Handbook of Computer-Communications Standards: Local Network Standards.* New York: Macmillan, vol. 2, 1987.

[11] W. Bux, "Performance Issues in Local Area Networks," *IBM Systems Jour.*, vol. 23. no. 4, 1984, pp. 351–374.

[12] A. S. Tanenbaum, *Computer Networks.* Englewood Cliffs, NJ: Prentice-Hall, 1981, pp. 286–320.

[13] W. Stallings, "Local Network Performance," *IEEE Comm. Mag.*, vol. 22, no. 2, Feb. 1984, pp. 27–36.

[14] R. Gail and L. Kleinrock, "An Invariant Property of Computer Network Power," *IEEE '81 ICC,* (4 vols.), July 1981, pp. 63.1.1–63.1.5.

Problems

1.1 Discuss the three common topologies suitable for LANs. Mention the merits and demerits of each topology.

1.2 Compare the relative advantages and disadvantages of token-passing ring and token-passing bus topologies.

1.3 Find out which LAN your campus or company has. Describe its topology, access mechanism, and transmission medium.

1.4 Is a network layer (OSI layer 3) needed in a broadcast network? Explain.

1.5 In a token ring LAN, which functions are performed by the network layer of the OSI model?

1.6 List the common factors that affect the performance of a LAN.

1.7 For a token ring LAN with a data rate of 1 Mbps, packet length of 1,000 bits, and token length of 24 bits, calculate the throughput and utilization.

1.8 Equation (1.7) applies to token ring or token bus. For CSMA/CD,

$$U = \frac{1}{1 + [2a(1 - A)/A]}$$

where $A = (1 - 1/N)^{N-1}$ = the probability that exactly one station attempts a transmission in a slot and therefore acquires the medium. Find U when the number N of active stations is very large.

1.9 In a Cambridge ring with a data rate of 5 Mbps, each slot has 37 bits. If 50 stations are connected to the ring and the average internodal distance is 20 m, how many slots can the ring carry? Assume a propagation speed of 2.5×10^8 m/s and that there is a 1-bit delay at each station.

Chapter 2

Analytical Models of LAN

We sow a thought and reap an act. We sow an act and reap a habit. We sow a habit and reap a character. We sow a character and reap a destiny.

—William M. Thackeray

2.1 Introduction

When designing a local area network (LAN), establishing the performance characteristics of a network before putting it into use is of paramount importance; it gives the designer the freedom and flexibility to adjust various parameters of the network in the planning rather than the operational phase. However, it is hard to predict the performance of the LAN unless a detailed analysis of a similar network is available. Information on a similar network is generally hard to come by, so performance modeling of the LAN must be carried out. The three commonly used prediction methods are [1, 2]

- Analytic models
- Simulation models
- Hybrid models

Analytic models are mathematical representations of the system through which the system input and output variables are related. To make the mathematics involved tractable, several assumptions about the system are usually made. These assumptions tend to make the models unlike the real-life systems and lead to inaccurate results. Analytic models show qualitative relationships between input and output parameters in a better way than other techniques do. They are usually more difficult to develop than simulation models. Analytic models are useful when gross answers are acceptable and for rapid initial assessment.

Simulation models are usually computer programs to relate input and output variables of the system. Using simulation, a network may be modeled to any desired level of detail if the necessary system relationships are known. Simulation models can give more accurate results than analytical models because most of the assumptions made in the analytic models can be relaxed. However, the level of detail dictates the simulation run time. This makes detailed simulation slow and expensive.

Hybrid models combine the strong points of both analytic and simulation models. It is difficult to model an entire system in detail. The hybrid model is a compromise that offers the flexibility and speed of analytical models and the accuracy of the simulation models.

In this chapter we focus on the analytic models of four important protocols: the token-passing access methods for the ring and bus topologies, the CSMA/CD for bus, and the star. Not only will the analytic models offer a way of checking the simulation models to be discussed in later chapters, they also provide an insight into the nature of the networks we shall be simulating.

The primary performance criterion is the delay-throughput characteristics of the system. The mean transfer delay of a message is the time interval between the instant the message is available at the sending station and the end of its successful reception at the receiving station. It is convenient to regard the transfer delay as consisting of three components. The first component, W, is called the waiting time or access time. It is the time elapsed from the availability of a message in the source station transmit buffer until the beginning of its transmission on the channel. The second component, T_p, called the propagation time, is the time elapsed from the beginning of the transmission of the message until the arrival of the first bit of the message at the destination. The third component is the transmission or service time, S, which is the time elapsed between the arrival of the first bit of the message at the destination and the arrival of the last bit. As soon as the last bit arrives at the destination, the transfer is complete. This implies that the transfer delay D includes the waiting time W (or queueing delay) at the sending station, the service (or transmission) time S of the message, and the propagation delay T_p; that is,

$$D = W + S + T_p \tag{2.1a}$$

In terms of their expected values

$$E(D) = E(W) + E(S) + E(T_p) \tag{2.1b}$$

2.2 Token-Passing Ring

The token-passing ring, developed at the Zurich Research Laboratories of IBM in 1972 and standardized as an access method in the IEEE Standard 802.5, is the best-known of all the ring systems. Here we are interested in its basic operation and delay analysis [3, 4].

2.2.1 Basic Operation

In a token ring, the stations are connected as in all ring networks as illustrated in Figure 2.1. Access to the transmission channel is controlled by means of a special eight-bit pattern called a *token*, which is passed around the ring. When the system is initialized, a designated station generates a free token, such as 11111111. If no station is ready to transmit, the free token circulates around the ring. When a station wishes to transmit, it captures the free token and changes it to a busy token, such as 11111110, thereby disallowing other stations from transmitting. The packet to be transmitted is appended to the busy token. The receiving station copies the information. When the information reaches the sending station, the station takes it off the ring and generates a new free token to be used by another station that may need the transmission channel.

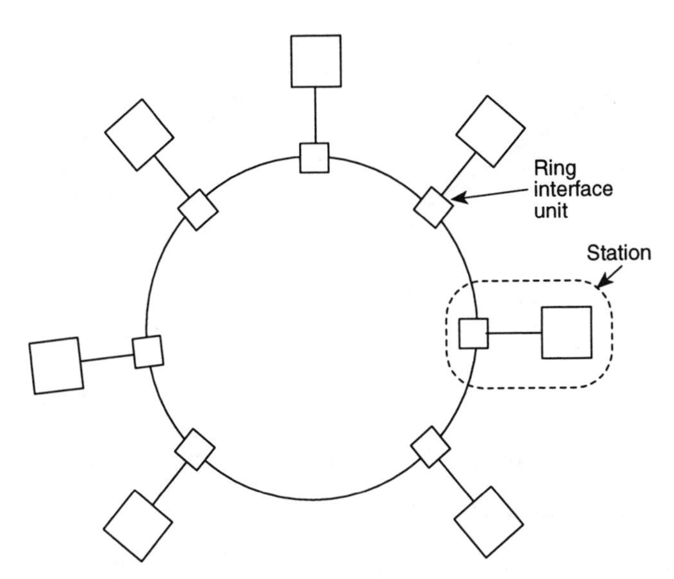

Figure 2.1 A typical ring topology. From J. L. Hammond and P.J.P. O'Reilly, *Performance Analysis of Local Computer Networks*, 1986, by permission of Addison-Wesley.

This operation can be described by a single-server queueing model, as illustrated in Figure 2.2. The server serves as many queues as stations attached to the ring. The server attends the queues in a cyclic order as shown by the rotating switch, which represents the free token. Once a station captures the token, it is served according to one of the following service disciplines:

- Exhaustive service: The server serves a queue until there are no customers left in that queue.
- Gated service: The server serves only those customers in a queue that were waiting when it arrived at that queue (i.e., when the server arrives at a queue, a gate is closed behind the waiting customers and only those customers in front of the gate are served).
- Limited service: The server serves a limited number of customers, say k (constant) or fewer, that were waiting when it arrived at the queue.

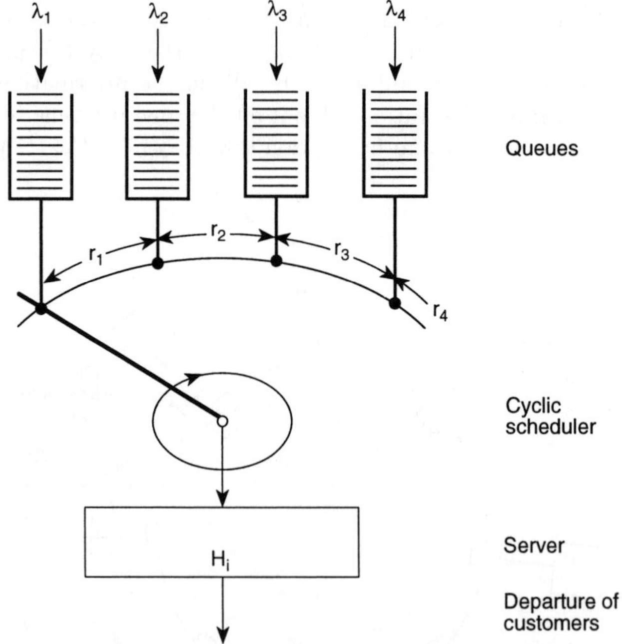

Figure 2.2 Cyclic-service queueing model.

2.2.2 Delay Analysis

Consider a single server serving N queues in a cyclic manner as shown in Figure 2.2. Let r_i denote a constant switchover time from queue i to queue $i + 1$ and R_o be the sum of all switchover times—i.e.,

$$R_o = \sum_{i=1}^{N} r_i \tag{2.2}$$

We examine the M/G/1 model; that is, messages arrive at queues according to independent Poisson processes with mean rates $\lambda_1, \lambda_2, \cdots, \lambda_N$ and the service times H_i of the messages from queue i are generally distributed with mean $E(S_i)$ and second moment $E(S_i^2)$. We denote the utilization of queue i by

$$\rho_i = \lambda_i E(S_i) \tag{2.3}$$

and assume that the normalization condition

$$\rho = \sum_{i=1}^{N} \rho_i < 1 \tag{2.4}$$

holds. Let V_i denote the intervisit time of queue i, also known as the server-vacation time, the time interval from the server's departure from the queue until its return to the same queue. The moment generating function for the statistical-equilibrium waiting time distribution is given by [5–7]:
Exhaustive service:

$$G_W^e(z) = E(e^{-zW_i}) = \frac{1-\rho_i}{E(V_i)} \frac{1 - G_v(z)}{z - \lambda_i + \lambda_i G_s(z)} \tag{2.5}$$

Gated service:

$$G_W^g(z) = \frac{G_c \lambda_i [1 - G_s(z)] - G_c(z)}{E(V_i)[z - \lambda_i + \lambda_i G_s(z)]} \tag{2.6}$$

Limited service:

$$G_W^\ell(z) = \frac{1 - \rho_i + \lambda_i E(V_i)}{E(V_i)} \frac{1 - G_v(z)}{z - \lambda_i + \lambda_i G_s(z) G_v(s)} \tag{2.7}$$

where $G_v(z) = E(e^{-zV_i})$ is the generating function for the intervisit time, $G_s(z) = E(e^{-zS_i})$ is the generating function for the service time, and $G_c(z) = E(e^{-zC_i})$ is the generating function for the cycle time.

From Eqs. (2.5) to (2.7), the mean waiting time of messages in queue i is determined by differentiating $G_W(z)$ and setting $z = 0$. The result is:
Exhaustive service:

$$E^e(W_i) = \frac{E(V_i)}{2} + \frac{Var(V_i)}{2E(V_i)} + \frac{\lambda_i E(S_i^2)}{2(1-\rho_i)} \tag{2.8}$$

Gated service:

$$E^g(W_i) = \frac{E(C_i)}{2} + \frac{Var(C_i)}{2E(C_i)} + \frac{\rho_i E(S_i^2)}{2(1-\rho_i)E(S_i)} \tag{2.9}$$

Limited service:

$$E^{\ell}(W_i) = \frac{\lambda_i E[(V_i + S_i)^2]}{2[1 - \rho_i + \lambda_i E(V_i)]} \tag{2.10}$$

Hence the mean waiting time can be found provided the first two moments of the intervisit times V_i are known.

To find the first moment of V_i, let C_i be the total cycle time (i.e. the time between subsequent visits of the server to queue i) and T_i be the time spent by the server at queue i, then

$$E(V_i) = E(C_i) - E(T_i) \tag{2.11}$$

It is readily shown that [8]

$$E(C_i) = \frac{R_o}{1 - \rho} \tag{2.12}$$

Since the traffic flow must be conserved, the average number of messages serviced during one visit of queue i is equal to the average number of arriving messages at that queue in one cycle time; i.e.,

$$\frac{E(T_i)}{E(S_i)} = \lambda_i E(C_i)$$

or

$$E(T_i) = \lambda_i E(C_i) E(S_i) \tag{2.13}$$

Substituting Eqs. (2.12) and (2.13) into Eq. (2.11) gives the mean intervisit time of queue i as

$$E(V_i) = \frac{1 - \rho_i}{1 - \rho} R_o \tag{2.14}$$

Introducing Eqs. (2.12) and (2.14) in Eq. (2.8) leads to

$$E^{e}(W_i) = \frac{Var(V_i)}{2E(V_i)} + \frac{1 - \rho_i}{2(1 - \rho)} R_o + \frac{\rho_i}{2(1 - \rho_i)} \frac{E(S_i^2)}{E(S_i)} \tag{2.15}$$

for exhaustive service. A similar procedure for gated service discipline results in

$$E^{g}(W_i) = \frac{Var(V_i)}{2E(V_i)} + \frac{1 + \rho_i}{2(1 - \rho)} R_o + \frac{\rho_i}{2(1 - \rho_i)} \frac{E(S_i^2)}{E(S_i)} \tag{2.16}$$

For a limited-service, continuous-time system, we have an explicit solution for $E(W_i)$ only in the special case of statistically symmetric conditions and $k = 1$ for all stations [5, 7]. However, an upper bound for $E(W_i)$ for any k has been determined [9].

For symmetric traffic conditions (i.e., in the case of identical stations),

$$\lambda_1 = \lambda_2 = \ldots = \lambda_N = \frac{\lambda}{N} \tag{2.17}$$

$$r_1 = r_2 = \ldots = r_N = R_o/N = r \tag{2.18}$$

and the mean waiting time for all the queues becomes
Exhaustive service:

$$E^e(W_i) = \frac{\delta^2}{2r} + \frac{Nr(1-\rho/N)}{2(1-\rho)} + \frac{\rho E(S^2)}{2(1-\rho)E(S)} \tag{2.19}$$

Gated service:

$$E^g(W_i) = \frac{\delta^2}{2r} + \frac{Nr(1+\rho/N)}{2(1-\rho)} + \frac{\rho E(S^2)}{2(1-\rho)E(S)} \tag{2.20}$$

Limited service:

$$E^\ell(W_i) = \frac{\delta^2}{2r} + \frac{Nr(1+\rho/N)+N\lambda\delta^2}{2(1-\rho-N\lambda r)} + \frac{\rho E(S^2)}{2(1-\rho-N\lambda r)E(S)} \tag{2.21}$$

where δ^2 is the variance of the switchover time. An alternative, less rigorous means of deriving Eqs.(2.19) to (2.21) is the decomposition theorem [8]. Note that the only difference between Eqs.(2.19) and (2.20) is the sign in the term $(1 \pm \rho/N)$, which implies that $E^e(W) \leq E^g(W)$. Thus, from Eqs. (2.19) to (2.21), we conclude that

$$E^e(W) \leq E^g(W) \leq E^\ell(W) \tag{2.22}$$

These derivations are for continuous-time systems. Corresponding derivations for discrete-time systems have also been found [5, 9–11].

The formulas in Eqs.(2.19) to (2.21) for the waiting time are valid for token ring and token bus protocols. However, the mean value r of the switchover time and its variance δ^2 differ for each protocol. Here we evaluate these parameters for the token ring.

The token-passing interval or switchover time T is given by

$$T = T_t + T_{pt} + T_b \tag{2.23}$$

where T_t is the token transmission time, T_{pt} is the token propagation delay, and T_b is the bit delay per station. Hence, the expected value $r = E(T)$ is given by

$$r = E(T_t) + E(T_{pt}) + E(T_b) \tag{2.24}$$

and, since the random variables are independent, the variance $Var(T) = \delta^2$ is given by

$$\delta^2 = Var(T_t) + Var(T_{pt}) + Var(T_b) \tag{2.25}$$

Assuming a constant token packet length L_t (including preamble bits), for a network data rate R,

$$T_t = \frac{L_t}{R}$$

Its expected value is constant. Hence

$$E(T_t) = T_t = \frac{L_t}{R}, \qquad Var(T_t) = 0 \tag{2.26}$$

Assuming that the stations are equally spaced on the ring, the distance between any adjacent stations is identical to ℓ/N, where ℓ is the physical length of the ring. If P is the signal propagation delay in seconds per unit length (the reciprocal of the signal propagation delay velocity u, i.e., $P = 1/u$), the token propagation delay is

$$T_{pt} = \frac{P\ell}{N}$$

Hence

$$E(T_{pt}) = T_{pt} = \frac{P\ell}{N} \qquad Var(T_{pt}) = 0 \tag{2.27}$$

If L_b is the bit delay caused by each station,

$$T_b = \frac{L_b}{R}$$

and

$$E(T_b) = \frac{L_b}{R} \qquad Var(T_b) = 0 \tag{2.28}$$

We conclude from Eqs. (2.24) to (2.28) that

$$r = P\ell/N + (L_b + L_t)/R \qquad \delta^2 = 0 \tag{2.29}$$

The average propagation suffered from one station is the propagation delay halfway around the ring: i.e.,

$$E(T_p) = \tau/2 \tag{2.30}$$

where τ is the round-trip propagation delay. Note that the sum of the switchover times (assumed to be constant) corresponds to the round-trip propagation delay and the sum of the bit-holding times at each station; i.e.,

$$Nr = P\ell + N(L_b + L_t)/R = \tau \tag{2.31}$$

Thus, for large N and symmetric traffic conditions, the mean transfer delay is obtained by substituting Eqs. (2.19), (2.20), (2.21), (2.29), and (2.30) in Eq. (2.1). We obtain
Exhaustive service:

$$E^e(D) = \frac{\tau(1-\rho/N)}{2(1-\rho)} + \frac{\rho E(S^2)}{2(1-\rho)E(S)} + E(S) + \tau/2 \tag{2.32}$$

Gated service:

$$E^g(D) = \frac{\tau(1+\rho/N)}{2(1-\rho)} + \frac{\rho E(S^2)}{2(1-\rho)E(S)} + E(S) + \tau/2 \tag{2.33}$$

Limited service:

$$E^\ell(D) = \frac{\tau(1+\rho/N)}{2(1-\rho-\lambda\tau)} + \frac{\rho E(S^2)}{2(1-\rho-\lambda\tau)E(S)} + E(S) + \tau/2 \tag{2.34}$$

Finally, the mean service time $E(S)$ is given by

$$E(S) = (E(L_p) + L_h)/R = \rho/\lambda \tag{2.35a}$$

where L_p and L_h are the mean packet length and header length. For fixed messages (requiring constant service times),

$$E(S^2) = E^2(S) \tag{2.35b}$$

and for exponential service times,

$$E(S^2) = 2E^2(S) \tag{2.35c}$$

Example 2.1

Messages arrive at a swiching node at the rate of 2 bits/minute, as shown in Figure 2.3. If the message is exponentially distributed with an average length of 20 bytes and the node serves 10 bits/second, calculate the traffic intensity.

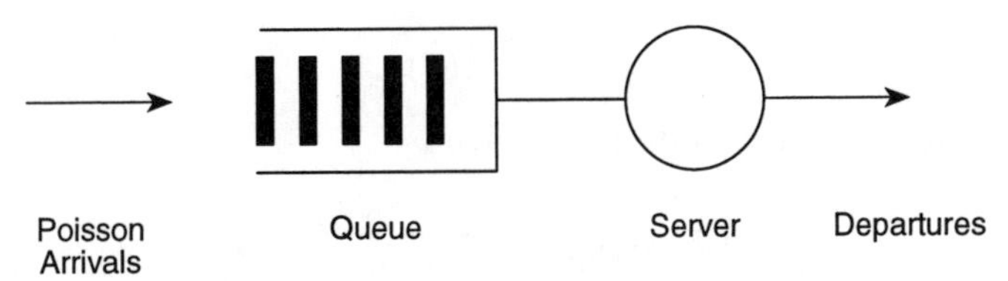

Figure 2.3 A switching for Example 2.1.

Solution

The arrival rate is

$$\lambda = 2 \text{ bits/minute} = \frac{2}{60} \text{ bps}$$

The service time is

$$E(S) = \frac{L_p}{R} = \frac{20 \times 8}{10} = 16 \text{ s}$$

The traffic intensity is given by

$$\rho = \lambda E(S) = \frac{2}{60} \times 16 = 0.5333$$

Example 2.2

A token-ring LAN has a total propagation delay of 20μs, a channel capacity of 10^6 bps, and 50 stations, each of which generates Poisson traffic and has a latency of 1 bit. For a traffic intensity of 0.6, calculate

(a) The switchover time
(b) The mean service time
(c) The message arrival rate per station
(d) The average delay for exhaustive, gated, and limited service disciplines

Assume 10 bits for overhead and 500 bits average packet length, exponentially distributed.

Solution

(a) If the end-to-end propagation time is $\tau = 20\mu$s, then the switchover time r is given by

$$r = \frac{\tau}{N} = \frac{20}{50} = 0.4 \ \mu\text{s}$$

(b) The mean service time is

$$E(S) = \frac{E(L_p) + L_h}{R} = \frac{500 + 10}{10^6} = 510 \ \mu\text{s}$$

(c) Since $\rho = \lambda E(S)$, the total arrival rate is

$$\lambda = \frac{\rho}{E(S)}$$

Hence the arrival rate per station is

$$\lambda_i = \frac{\rho}{NE(S)} = \frac{0.6}{50 \times 510 \times 10^{-6}}$$
$$= 23.53 \text{ bps}$$

(d) For exponentially distributed packet lengths,

$$E(S^2) = 2E^2(S) = 52.02 \times 10^{-8} \text{ s}^2$$

Using Eqs. (2.32) to (2.34), we obtain

$$\begin{aligned} E^e(D) &= \frac{20 \times 10^{-6}(1 - 0.6/50)}{2(1-0.6)} + \frac{0.6 \times 52.02 \times 10^{-8}}{2(1-0.6) \times 510 \times 10^{-6}} \\ &+ 510 \times 10^{-6} + 10 \times 10^{-6} \\ &= (24.7 + 765 + 520)\ \mu\text{s} = 1.3097 \text{ ms} \end{aligned}$$

for exhaustive service. For gated service,

$$\begin{aligned} E^g(D) &= \frac{20 \times 10^{-6}(1 + 0.6/50)}{2(1-0.6)} + (765 + 520)\ \mu\text{s} \\ &= 1.3103 \text{ ms} \end{aligned}$$

For limited service,

$$\begin{aligned} E^\ell(D) &= \frac{20 \times 10^{-6}(1 + 0.6/50)}{2(1-0.6-0.02353)} + \frac{0.6 \times 52.02 \times 10^{-8}}{2(1-0.6-0.02353) \times 510 \times 10^{-6}} \\ &+ 510 \times 10^{-6} + 10 \times 10^{-6} \\ &= (26.881 + 812.81 + 520)\ \mu\text{s} = 1.3597 \text{ ms} \end{aligned}$$

Notice that

$$E^e(D) < E^g(D) < E^\ell(D)$$

as suggested in Eq. (2.22).

2.3 Token-Passing Bus

The token bus protocol was inspired by the token ring and standardized in the IEEE Standard 802.4. The basic operation of the token bus LAN is fully discussed by Hammond and O'Reilly [3] and Kauffels [12], and its delay analysis is presented by Sachs and colleagues [13].

2.3.1 Basic Operation

The operation of the token bus protocol is similar in many respects to that of the token ring. Although the token bus uses bus topology whereas the token ring uses ring topology, the stations on a token bus are logically ordered to form a *logical ring*, which is not necessarily the same as the physical ordering of the stations. Figure 2.4 shows a typical ordering of stations on a bus with the sequence AEFHCBA. Each station on the ring knows the identity of the stations preceding and following it. The right of access to the bus is controlled by the cyclic passing of a token among the stations in the logical ring. Unlike in a token ring where the token is passed implicitly, an explicit token with node addressing information is used. The token is passed in order of address. When a station receives the token, it may transmit its messages according to a service discipline (exhaustive, gated, or limited) and pass the token to the next station in the logical ring.

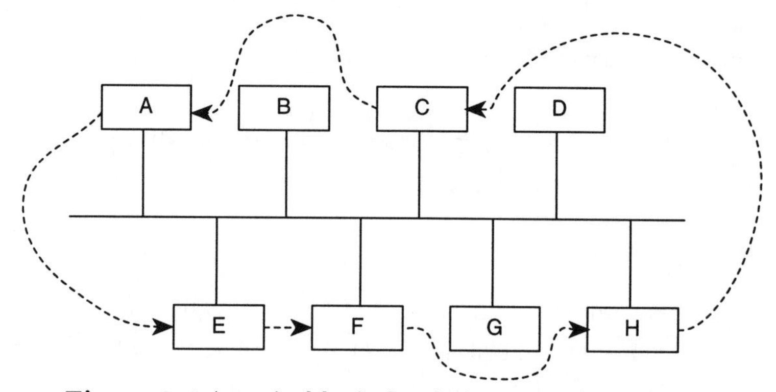

Figure 2.4 A typical logical ordering on a physical bus.

A token bus protocol differs in some respects from a token ring. Since token bus is a broadcast protocol, stations not in the logical ring can receive messages. The interfaces of stations on a token bus are passive and thus create no station latency or delay, unlike the token ring, where the signal is regenerated at each station. Propagation delays on a token bus are generally longer because the token may have to travel longer distances to satisfy the logical ordering of the stations.

2.3.2 Delay Analysis

As mentioned earlier, the expressions for waiting time (or queueing delay) in Eqs. (2.19) to (2.21) are valid for both token ring and token bus protocols except that the mean value of r of the switchover time and its variance δ^2 are different for the two protocols. We now evaluate these parameters as they apply to the token bus.

Unlike the token ring protocol, the token bus protocol requires that the complete token be transmitted, received, and identified before a data packet can be generated and transmitted. Therefore, the token passing transmission time T_t is a significant delay in the token bus protocol. According to Eq. (2.26),

$$E(T_t) = T_t = \frac{L_t}{R}, \qquad Var(T_t) = 0 \tag{2.36}$$

Assuming bus length ℓ, uniform distribution of N stations, and an equal probability of communication between any two stations, the distance between station i and its logical successor j is given by

$$d_{ij} = jd = \frac{j\ell}{N-1}, \quad 1 \le j \le N-1 \tag{2.37}$$

The probability of station i having the token and passing it to station j is given by

$$P_{ij} = \frac{1}{\binom{N}{2}} = \frac{2}{N(N-1)} \tag{2.38}$$

If X is the token propagation distance, the expected token propagation delay is

$$\begin{aligned} E(X) &= \sum d_{ij} P_{ij} = \sum_{i=1}^{N} \sum_{j=1}^{i-1} \frac{2\ell j}{N(N-1)^2} \\ &= \frac{(N+1)\ell}{3(N-1)} \end{aligned} \tag{2.39}$$

where the identities

$$\sum_{i=1}^{n} i = \frac{n}{2}(n+1)$$

and

$$\sum_{i=1}^{n} i^2 = \frac{n}{6}(n+1)(2n+1)$$

have been applied. Corresponding to the bus length ℓ, we have an end-to-end propagation delay τ. Therefore, the expected token propagation delay is

$$E(T_{pt}) = \frac{(N+1)\tau}{3(N-1)} \tag{2.40}$$

The variance of X is given by

$$\begin{aligned} Var(X) &= E(X^2) + [E(X)]^2 = \sum d_{ij}^2 P_{ij} - [E(X)]^2 \\ &= \frac{(N+1)\ell^2}{3(N-1)^3} - \frac{(N+1)^2\ell^2}{9(N-1)^2} \end{aligned} \tag{2.41}$$

where the identity

$$\sum_{i=1}^{n} i^3 = \left[\sum_{i=1}^{n} i\right]^2 = \frac{n^2}{4}(n+1)^2$$

has been incorporated. Thus the variance of the token passing propagation delay is

$$Var(T_{pt}) = \frac{(N+1)(N-2)\tau^2}{18(N-1)^2} \tag{2.42}$$

The bit delay per station adds to the token-passing time a delay corresponding to token handling and address recognition. In IEEE 802.4, for example, a buffer of four or five bits may be required depending on the size of the address field. If L_b is the bit delay caused by each station,

$$T_b = \frac{L_b}{R}.$$

Hence

$$E(T_b) = \frac{L_b}{R}, \qquad Var(T_b) = 0 \tag{2.43}$$

Substitution of Eqs. (2.32), (2.36), (2.42), and (2.43) into Eq. (2.24) yields

$$r = \frac{(N+1)\tau}{3(N-1)} + c, \qquad \delta^2 = \frac{(N+1)(N-2)\tau^2}{18(N-1)^2} \tag{2.44}$$

with limiting values ($N \to \infty$) of

$$r = \frac{\tau}{3} + c, \qquad \delta^2 = \frac{\tau^2}{18} \tag{2.45}$$

where

$$c = T_t + T_b = \frac{L_t + L_b}{R}$$

The packet propagation delay is the same as the token propagation delay, so for large N,

$$E(T_p) = \tau/3 \tag{2.46}$$

If we assume large N and symmetric traffic conditions, the mean transfer time is obtained by substituting Eqs. (2.19), (2.20), (2.21), (2.45), and (2.46) into Eq. (2.1).
Exhaustive service:

$$E^e(D) = \frac{\tau^2}{36(\tau/3+c)} + N(\tau/3+c)\frac{(1-\rho/N)}{2(1-\rho)} + \frac{\rho E(S^2)}{2(1-\rho)E(S)} + E(S) + \tau/3 \tag{2.47}$$

Gated service:

$$E^g(D) = \frac{\tau^2}{36(\tau/3+c)} + N(\tau/3+c)\frac{(1+\rho/N)}{2(1-\rho)} + \frac{\rho E(S^2)}{2(1-\rho)E(S)} + E(S) + \tau/3 \tag{2.48}$$

Limited service:

$$E^{\ell}(D) = \frac{\tau^2}{36(\tau/3+c)} + \frac{N(\tau/3+c)(1+\rho/N)}{2[1-\rho-N\lambda(\tau/3+c)]} + \frac{\rho E(S^2)}{2[1-\rho-N\lambda(\tau/3+c)]E(S)} + E(S) + \tau/3 \tag{2.49}$$

where the mean service time $E(S)$ is given by Eq. (2.35) and the end-to-end propagation delay by

$$\tau = P\ell \tag{2.50}$$

2.4 CSMA/CD Bus

Multiple access local area network (LAN) protocols divide broadly into two classes [14]: *random* (or *contention*) *access* protocols and *controlled access* protocols. In random access protocols, transmission rights are simultaneously offered to a group of stations in the hope that exactly one of the stations has a packet to send. However, if two or more stations send packets on the channel simultaneously, these messages interfere with each other and none of them are correctly received by the destination stations. In such cases, a collision has occured and stations retransmit packets until they are successfully received by the destination stations.

A controlled access mechanism is one in which a token is first secured by a node in order to transmit its messages through the medium. Controlled access protocols, such as the token ring and token bus considered in Sections 2.2 and 2.3 respectively, avoid collisions by coordinating access of the stations to the channel by imposing either a predetermined or dynamically determined order of access. Access coordination is done by use of the channel itself. Each station indicates with a short message on the channel whether or not it wants access. This polling mechanism consumes some channel capacity regardless of whether stations require access or not. Although such protocols are efficient when traffic is heavy, under light traffic conditions they result in unnecessary packet delays as stations that want to transmit wait their turn.

In contrast, random access protocols exhibit small packet delays under light traffic conditions: stations transmit as soon as they want access to the channel, and the probability of a collision is low when traffic is light. Another attractive aspect of random access protocols is their simplicity, making them easy to implement at the stations [15].

The Aloha family of protocols is popular because it was the first random access mechanism to be introduced. In this type of protocol, the success of a transmission is not guaranteed in advance. When two or more packets overlap in time, even by a bit, all are lost and must be retransmitted. The carrier sense multiple access (CSMA) reduces the level of interference caused by overlapping packets by allowing users to sense the carrier due to other users' transmissions and aborting transmission when the channel is sensed busy. In CSMA, all nodes listen constantly to the bus and only transmit if there is no transmission already of the bus. This is the *carrier sense* aspect of the name. If there is no transmission on the bus, any node with available data can transmit immediately, hence the term *multiple access.* Besides the ability to sense the carrier, some LANs are also able to detect interference among several transmissions while transmitting and to abort transmission when there is collision. This additional feature produces a variation of CSMA that is known as CSMA/CD (Carrier Sense Multiple Access with Collision Detection). Because of its simplicity, CSMA/CD is perhaps the most popular contention-based protocol. It operates on a bus-type network and is sometimes referred to as the Ethernet protocol.

2.4.1 Basic Operation

In a LAN employing the CSMA/CD protocol, each node listens during, as well as before, transmission its packet. Variations within the CSMA/CD protocols center about the operation mode of the station when the medium is sensed busy or idle. The most popular operation modes are [15, 16]

- Nonpersistent
- One-persistent
- p-Persistent

In the nonpersistent CSMA/CD scheme, a node with a packet ready for transmission senses the channel and acts as follows:

1. If the channel is sensed idle, the node initiates transmission of the packet.
2. If the channel is sensed busy, the node schedules the retransmission of its packet to some later time. It waits for a random amount of time and resenses the channel.
3. If a collision is detected during transmission, the node aborts its transmission and schedules the retransmission of the packet later.

In the 1-persistent CSMA/CD protocol (which is a special case of the p-persisent), a node that finds the channel busy persists in transmitting as soon as the channel becomes free. If it finds the channel idle, it transmits the packet immediately with probability one. In other words, a ready node senses the channel and proceeds as in nonpersistent CSMA/CD, except that, when the channel is sensed busy, the node monitors the channel until it is sensed idle and then with probability one initiates transmission of its packet.

In the p-persistent protocol, a ready node senses the channel and proceeds as in the nonpersistent protocol except that when the channel is sensed busy, the node persists until the channel is idle

i. With probability p it initiates transmission of the packet.
ii. With probability $1 - p$ it delays transmission by τ seconds (the end-to-end propagation delay).

If at this instant the channel is sensed idle, then the node repeats steps (i) and (ii); otherwise it schedules retransmission of its packet later.

Note that in all CSMA/CD protocols, given that a transmission is initiated on an empty channel, it takes at most τ seconds for the packet transmission to reach all nodes. Beyond this time the channel will surely be sensed busy for as long as data transmission is in process. A collision can only occur if another transmission is initiated before the current one is sensed, and it will take at most additional τ seconds before interference reaches all devices. Moreover, Ethernet has a collision consensus reinforcement mechanism by which a device, experiencing interference, jams the channel to ensure that all other interfering nodes detect the collision.

In addition to the variations in the protocols, the transmission medium may be slotted or unslotted.

2.4.2 Delay Analysis

A widely used analytic model of CSMA/CD networks was developed by Lam [17, 18]. The analysis of the M/G/1 queue using embedded Markov chains led to a closed-form expression for the mean delay $E(D)$. The underlying assumptions are close to the standardized CSMA/CD protocol, and the results are simple to evaluate numerically.

The underlying assumptions in Lam's model are as follows. The network consists of an infinite number of stations connected to a slotted channel in which stations can begin transmissions only at the start of a time slot. The traffic offered to the network is a Poisson process with a constant arrival rate λ. Each state is allowed to hold at most one message at a time. Message transmission times are generally distributed. The system operates under the p-persistent protocol. Following a successful transmission, all ready stations transmit within the next slot. Following a collision, stations use an adaptive retransmission algorithm such that the probability of a successful transmission within any of the slots subsequent to a collision is constant and equal to $1/e$ $(= 0.368)$.

Under these assumptions, the mean delay was found by Lam and later modified by Bux [4,19] for a nonslotted channel as:

$$E(D) = \frac{\lambda\left[E(S^2) + (4e+2)\tau E(S) + 5\tau^2 + 4e(2e-1)\tau^2\right]}{2\Big(1 - \lambda\left[E(S) + \tau + 2e\tau\right]\Big)}$$

$$-\frac{(1-e^{-2\lambda\tau})(e+\lambda\tau-3\lambda\tau e)}{\lambda e\left[F(\lambda)e^{-(1+\lambda\tau)}+e^{-2\lambda\tau}-1\right]}$$

$$+2\tau e+E(S)+\tau/3 \tag{2.51}$$

where τ is the end-to-end propagation delay as in Eq. (2.50), and $E(S)$ and $E(S^2)$ are respectively the first and second moments of the message transmission (or service) time as given by Eq. (2.35). The term $\tau/3$ is the mean source-to-destination propagation time $E(T_p)$. It is heuristically taken as $\tau/2$ in other works, but we have used $\tau/3$ to be consistent with the derivation in Eq.(2.40) or (2.46). The function $F(\lambda)$ is the Laplace transform of the message transmission time distribution; that is,

$$F(\lambda)=\int_0^\infty f(t)e^{-\lambda t}\,dt \tag{2.52}$$

For constant message lengths,

$$F(\lambda)=e^{-\rho},\quad E(S^2)=E^2(S) \tag{2.53}$$

where $\rho=\lambda E(S)$. For exponentially distributed message lengths,

$$F(\lambda)=\frac{1}{1+\rho},\quad E(S^2)=2E^2(S) \tag{2.54}$$

It is important to note the two limiting cases of operation of CSMA/CD from Eq. (2.51). The mean delay becomes unbounded as the traffic intensity ρ approaches the maximum value of

$$\rho\text{max}=\frac{1}{1+(2e+1)a}=\frac{1}{1+6.44a} \tag{2.55}$$

Also as the traffic intensity ρ approaches zero, the mean delay approaches the minimum value of

$$E(D)_{\text{min}}=E(S)+\tau/3 \tag{2.56}$$

Example 2.3

A CSMA/CD network with a channel bit rate of 1 Mbps connects 40 stations on a 2-km cable. For fixed packet length of 1,000 bits, calculate the mean transfer delay. Assume propagation delay of 5 μs/km and an average arrival rate/station of 0.015 packets/second.

Solution

The mean service time is

$$E(S) = \frac{L_p}{R} = \frac{1,000}{10^6} = 10^{-3} \text{ s}$$

The mean arrival rate for each station is

$$\lambda_i = 0.015 \times 1,000 \text{ bits/s} = 15 \text{ bps}$$

Hence the total arrival rate is

$$\lambda = N\lambda_i = 40 \times 15 = 600 \text{ bps}$$

The traffic intensity is

$$\rho = \lambda E(S) = 10^{-3} \times 600 = 0.6$$

The end-to-end propagation delay is

$$\tau = \frac{\ell}{u} = \ell P = 2 \text{ km} \times 5 \ \mu\text{s/km} = 10 \ \mu\text{s}$$

For constant packet lengths,

$$F(\lambda) = e^{-\rho}, \quad E(S^2) = E^2(S) = 10^{-6}$$

Applying Eq. (2.51), we obtain the delay as

E(D)

$$= \frac{600\left\{10^{-6} + \left[(4e+2)\cdot 10^{-5} \times 10^{-3}\right] + (5 \times 10^{-10}) + \left[4e(2e-1) \times 10^{-10}\right]\right\}}{2\left\{1 - 600\left[(10^{-3} + 10^{-5} + (2e \cdot 10^{-5})\right]\right\}}$$

$$- \frac{\left(1 - e^{-2\times 6\times 10^{-3}}\right)\left[e + (6 \times 10^{-3}) - (3e \cdot 6 \times 10^{-3})\right]}{600e\left[e^{-0.6}e^{-(1+6\times 10^{-3})} + e^{-12\times 10^{-3}} - 1\right]}$$

$$+ 2e \times 10^{-5} + 10^{-3} + \frac{10^{-5}}{3}$$

$$= (761.35 - 103.87 + 1005.77) \ \mu\text{s}$$

$$= 1.663 \text{ ms}$$

2.5 Star

Due to their simplicity, the star networks evolved as the first controlled-topology networks. They are regarded as the oldest communication medium topologies because of their use in centralized telephone exchanges. As we shall see, the star topology has some disadvantages, which have led to its apparent unpopularity in local area networks. Although the control of traffic is distributed in both the bus and the ring topologies, it is concentrated in the star.

2.5.1 Basic Operation

A star topology usually consists of a primary node (hub) and secondary nodes (the nodes on the periphery). The primary node is the central node, which acts like a switch or traffic director. Communication between any two nodes is via circuit switching. When a peripheral node has data to transmit, it must first send a request to the central node, which establishes a dedicated path between the node and the destination node. All links must therefore be full duplex to allow two-way communication between the primary and secondary nodes as shown in Figure 2.5.

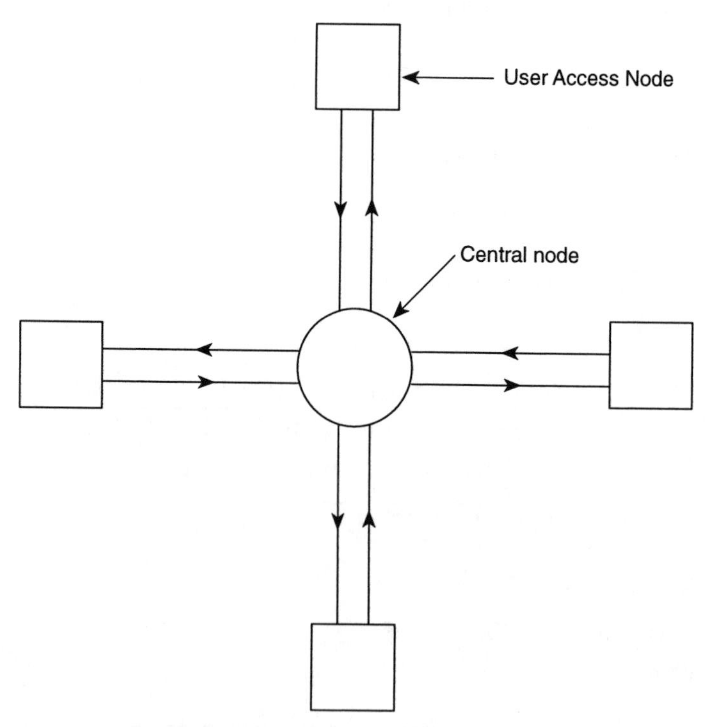

Figure 2.5 A typical star network.

The use of a central node to perform all routing provides a fairly good mapping of technology, but at the expense of creating a complex routing station. The central node is a complex one from a hardware standpoint. It is also a limiting element in the star's growth because it requires the hub to have a spare port to plug in a new link. The delay caused by the hub affects the performance of the network. Because of the problems associated with the central switch, the star network exhibits growth limitations, low reliability, poor expandability, and a complex central node. Although not as effective as the bus or ring in terms of routing, the star is used for other reasons.

The star networks offer positive features that many other networks lack. For example, the interconnection in star networks is point-to-point, which makes them suitable for optical fiber–based implementation. That is, in fiber-optic systems, star-shaped topologies are usually preferred because they allow the interconnection of more nodes, are less prone to castastrophic failure, and are relatively flexible and expandable. In fact, the first optical fiber networks were built in the star configuration [20]. Also, the throughput of star networks is usually very high and can easily approach unity, which means that the bandwidth is effectively utilized. Very high data rates can be sustained on star networks. Star systems allow simple modular expansion, and their performance is in general better than the performance of other networks [21].

2.5.2 Delay Analysis

Delay analyses of star networks have been carried out by Kamal [21] and Mehmet-Ali, Hayes, and Elhakeem [22]. Here we adopt the approximate analysis of the latter [22].

The underlying assumptions of the analysis are as follows: Messages are assumed to arrive at each source node according to a Poisson process with an average arrival rate of λ_i and have an arbitrary length distribution. Messages arrive to the system at one of the N nodes and are switched to one of the other $N-1$ nodes. It is assumed that the source-destination line pair must be free before a message can be transmitted and that the probability that a message will have its destination as its source is zero. It is also assumed that messages are transmitted from the source queues strictly in their order of arrival. Finally, it is assumed that the traffic is symmetric. With each source modeled as an $M/G/1$ queue, the waiting time or queueing delay is obtained as [22]:

$$E(W) = \hat{y} + \frac{\lambda \hat{y}^2}{2(1-\rho)} \tag{2.57}$$

where

$$\hat{y} = \big[1 + (N-2)\rho G\big]E(S) \tag{2.58a}$$

$$\hat{y}^2 = 2\big[1 + 2(N-2)\rho G + (N-2)(N-3)\rho^2 G^2\big]E(S^2) \tag{2.58b}$$

$$\rho = \frac{\lambda E(S)}{1-(N-2)G\lambda E(S)} \tag{2.58c}$$

$\lambda = \lambda_i$, and $G = 1/(N-1)$ is the probability that a message from source i will have node j as its destination. From Eq. (2.57), the stability requirement $\rho \leq 1$ implies that $\lambda E(S) \leq (N-1)(2N-3)$. For large N, this implies $\lambda E(S) \leq 1/2$.

The source–destination propagation time $E(T_p)$ is given by

$$E(T_p) = \tau \tag{2.59}$$

where τ is the round-trip or two-way propagation delay between any node and the central hub.

By substituting Eqs. (2.57) and (2.59) into Eq. (2.1), we obtain

$$E(D) = \hat{y} + \frac{\lambda \hat{y}^2}{2(1-\rho)} + E(S) + \tau \tag{2.60}$$

$E(S)$ and $E(S^2)$, the first and second moments of the message service time, are given by Eq. (2.35).

2.6 Performance Comparisons

Having examined each LAN protocol separately, it is instructive that we compare the protocols in terms of their performance under similar traffic conditions. We compare Eqs. (2.32), (2.47), (2.51), and (2.60) and present typical performance results for the four protocols. As expected, the components of the mean delay that depend on the propagation delay make a negligible contribution toward total delay. The queueing delays, on the other hand, contribute heavily to the total delay.

Figures 2.6 and 2.7 compare the delay/throughput characteristic of the four protocols. In both figures, the ordinate represents the mean delay normalized to the mean service time, $E(D)/E(S)$, while the abscissa denotes the traffic intensity or offered load, $\rho = \lambda E(S)$. In both figures, we consider:

N (no. of stations)	=	50
ℓ (cable length)	=	2 km
Packet length distribution	:	exponential
E(L_p) (mean packet length)	=	1,000 bits
L_h (header length)	=	24 bits
L_b (bit delay)	=	1 bit
L_t (token packet length)	=	0
P (propagation delay)	=	5 μs/km

Figure 2.6 shows the curves plotted for the four protocols when the transmission rate, R, is 1 Mb/s. It is apparent from Figure 2.6 that the star has the worst performance that the token ring performs less well than the token bus over the entire throughput range, and that the token bus and CSMA/CD protocols track one another closely over most of the throughput range.

Increasing the transmission rate to 10 Mb/s while keeping other parameters the same, we obtain the curves in Figure 2.7. It is evident from this figure that the performance of the star is still worst and the performance of both token-passing protocols is only slightly affected by the increased network rate, thus showing little sensitivity to this parameter. However, the CSMA/CD scheme is highly sensitive to the transmission rate. This should be expected because with an increase in the transmission rate, relatively more collisions take place and more transmission attempts result in collisions. In fact, 10 Mbps is the upper end for Ethernet (CSMA/CD), but this speed is generally unachievable due to a substantial increase in collisions. For this reason, 2 Mbps or less is typically taken as the upper achievable speed for CSMA/CD.

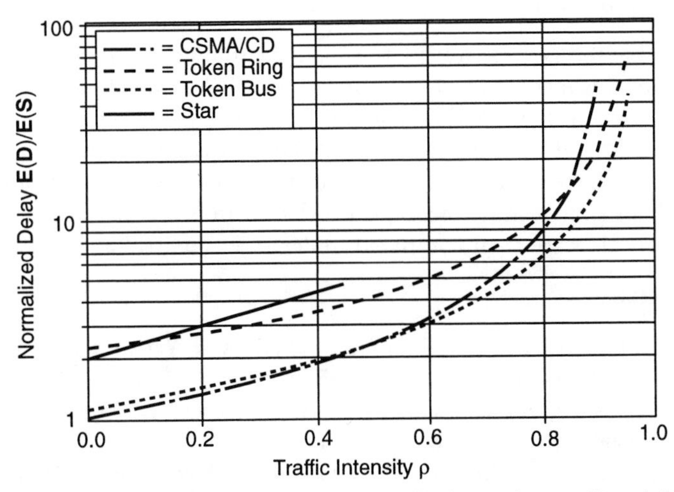

Figure 2.6 Normalized delay versus traffic intensity at $R = 1$ Mbps.

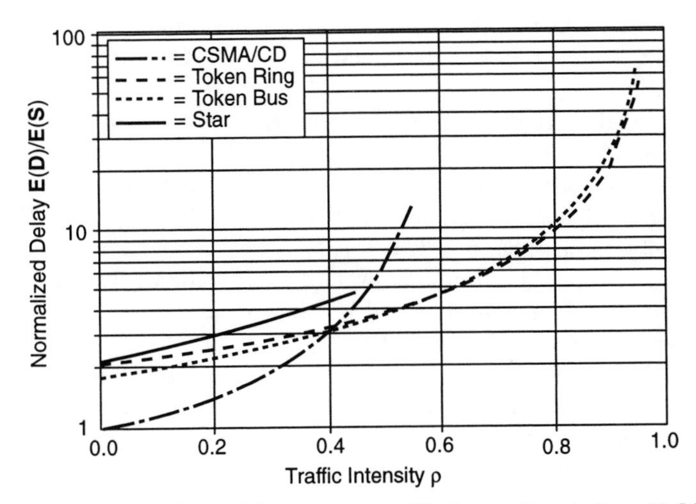

Figure 2.7 Normalized delay versus traffic intensity at $R = 10$ Mbps.

From performance grounds, CSMA/CD is better at light loading. For heavy loading, token ring seems to be more desirable than token bus, and certainly more desirable than CSMA/CD networks. As mentioned in Chapter 1, however, performance may not be the only consideration in selecting a LAN technology. From a reliability viewpoint, for example, token ring presents problems: whenever a station attached to the ring fails, the whole network fails because the message must be retransmitted at each station. Also, considering the ease of maintenance, availability, extendibility, and complexity of a physical layer design, a bus architecture has some advantages over a ring.

2.7 SUMMARY

In this chapter, we examined the delay/throughput characteristics of four local area networks: token ring, token bus, CSMA/CD bus, and star. In order to make a valid comparison between the schemes, we presented analytical models based on similar sets assumptions. Assuming an $M/G/1$ queueing model for each station in the network, we obtained closed form approximate formulas for the mean delay for each protocol. The protocols were then compared under the same traffic conditions.

References

[1] J. R. Freer, *Computer Communications and Networks.* New York: Plenum Press, 1988, pp. 284, 285.

[2] P. J. Fortier, *Handbook of LAN Technology.* New York: McGraw-Hill, 1992, chap. 16, pp. 377–385.

[3] J. L. Hammond and P. J. P. O'Reilly, *Performance Analysis of Local Computer Networks.* Reading, MA: Addison-Wesly, 1986, pp. 225–237.

[4] W. Bux, "Local-area subnetworks: a performance comparison," *IEEE Trans. Comm.*, vol. COM-29, no. 10, Oct. 1981, pp. 1465–1473.

[5] H. Tagaki, *Analysis of Polling Systems.* Cambridge, MA: MIT Press, 1986.

[6] M. J. Ferguson and Y. J. Aminetzah, "Exact results for nonsymmetric token ring systems," *IEEE Trans. Comm.*, vol. COM-33, no. 3, March 1985, pp. 223–231.

[7] K. S. Watson, "Performance evaluation of cyclic service strategies—a survey," in E. Gelenbe (ed.), *Performance '84.* Amster- dam: North-Holland, 1984, pp. 521–533.

[8] S. W. Fuhrmann and R. B. Cooper, "Application of decomposition principle in M/G/1 vacation model to two continuum cyclic queueing models—especially token-ring LANs," *AT&T Tech. Jour.*, vol. 64, no. 5, May/June 1985, pp. 1091–1099.

[9] I. Rubin and L. F. M. de Moraes, "Message delay analysis for polling and token multiple-access schemes for local communication networks," *IEEE Jour. Sel. Area Comm.*, vol. SAC-1, no. 5, Nov. 1983, pp. 935–947.

[10] A. G. Konheim and M. Reiser, "A queueing model with finite waiting room and blocking," *J. ACM*, vol. 23, no. 2, April 1976, pp. 328–341.

[11] G. B. Swartz, "Polling in a loop system," *J. ACM*, vol. 27, no. 1, Jan. 1980, pp. 42–59.

[12] F.-J. Kauffels, *Practical LANs Analysed.* Chichester, UK: Ellis Horwood, 1989.

[13] S. R. Sachs et al., "Performance analysis of a token-bus protocol and comparison with other LAN protocols," *Proc. 10th Conf. on Local Computer Networks*, Minneapolis, MN, Oct. 1985, pp. 46–51 (published by IEEE Computer Soc. Press, Silver Spring, MD).

[14] R. Rom and M. Sidi, *Multiple Access Protocols: Performance and Analysis.* New York: Springer-Verlag, 1990.

[15] G. E. Keiser, *Local Area Networks.* New York: McGraw-Hill, 1989.

[16] F. A. Tobagi and V. B. Hunt, "Performance analysis of carrier sense multiple access with collision detection," *Computer Networks*, vol. 4, 1980, pp. 245–259.

[17] S. S. Lam, "A carrier sense multiple access protocol for local networks," *Computer Networks*, vol. 4, no. 1, Jan. 1980, pp. 21–32.

[18] J. F. Hayes, *Modeling and Analysis of Computer Communications Networks.* New York: Plenum Press, 1984, pp. 226–230.

[19] W. Bux, "Performance issues in local-area networks," *IBM Syst. Jour.*, vol. 23, no. 4, 1984, pp. 351–374.

[20] E. S. Lee and P. I. P. Boulton, "The principles and performance of Hebnet: A 40 Mbits/s glass fiber local area network," *IEEE J. Select. Areas Comm.*, vol. SAC-1, Nov. 1983, pp. 711–720.

[21] A. E. Kamal, "Star local area networks: A performance study," *IEEE Trans. Comp.*, vol. C-36, no. 4, April 1987, pp. 484–499.

[22] M. K. Mehmet-Ali, J. F. Hayes, and A. K. Elhakeem, "Traffic analysis of a local area network with a star topology," *IEEE Trans. Comm.*, vol. COM-36, no. 6, June 1988, pp. 703–712.

Problems

2.1 Compare and contrast controlled access and random access protocols.

2.2 For a token-passing ring, assume the following parameters:

No. of stations	=	50
Transmission rate	=	1 Mbps
Mean packet length	=	1,000 bits (exponentially distributed)
Length of the ring	=	2 km
Token length	=	24 bits
Header length	=	0 bit
Bit delay	=	1 bit
Propagation delay	=	5 μs/km

Calculate the mean delay of a message for exhaustive service discipline for $\rho = 0.1, 0.2, \cdots, 0.9$.

2.3 For both constant and exponential packet distributions, calculate the mean delay for a token bus LAN with the following parameters:

No. of stations	=	50
Bit rate	=	5 Mbps
Mean packet length	=	1,000 bits
Bus length	=	1 km
Token length	=	96 bits
Header length	=	0 bit
Bit latency	=	1 bit
Propagation delay	=	5 μs/km

Try cases for $\rho = 0.1, 0.2, \cdots, 0.9$ and assume exhaustive service discipline.

2.4 Repeat Problem 2.3 for the CSMA/CD protocol.

2.5 (a) Assuming an exhaustive service discipline, calculate the average transfer delay of a token bus with the following parameters:

No. of stations	=	40
Transmission rate	=	1 Mbps
Mean packet length	=	500 bits (exponentially distributed)
Cable length	=	4 km
Token length	=	96 bits
Header length	=	0 bit
Bit delay	=	1 bit
Propagation delay	=	2 μs/km
Traffic intensity	=	0.4

(b) Repeat part (a) for a CSMA/CD bus LAN.

2.6 Rework Problem 2.5 for the case of a constant packet length of 1,000 bits.

2.7 Verify Eqs. (2.55) and (2.56).

Chapter 3

Simulation Models

What is maturity? It is being able to carry money without spending it; being able to bear an injustice without retaliation; being able to do one's duty even when one is not watched; being able to keep on the job until it is finished; being able to accept criticism without letting it whip one.

—Anonymous

3.1 Introduction

The first two chapters dealt with what we shall be simulating—local area computer networks. This chapter and the following one concentrate on the tools necessary for the simulation. In this chapter, we set the stage for the rest of the book by providing an overview of simulation: its historical background, importance, characteristics, and stages of development.

Simulation is a procedure in which one system is substituted for another system that it resembles in certain important aspects [1]. It can also be viewed as the act of performing experiments on a model of a given system. By a model, we mean a representation of the system under investigation.

Simulation emerged as a numerical problem-solving approach during World War II when the *Monte Carlo methods* were successfully used by John Von Neumann and Stanislaw Ulam of Los Alamos laboratory. The Monte Carlo methods were applied to problems related to the atomic bomb. Simulation was introduced into university curricula in the 1960s, when books and periodicals on simulation began to appear.

The term *Monte Carlo* is the name of a city that is well known for its gambling activities and has been associated with simulation that entails using random variables. Although some authors use the terms *Monte Carlo simulation* and *simulation* synonymously because they both involve the use of random variable, there is a distinction between the two. The system that is being modeled is deterministic in Monte Carlo simulation, and stochastic in the case of ordinary simulation [2].

Computer systems can be modeled at several levels of detail [3]: circuit level, gate level, and system level. At the circuit level, we employ simulation to analyze the switching behavior of various components of the circuit such as resistors, capacitors, and transitors. In the gate-level simulation, the circuit components are aggregated into a single element, so the element is analyzed from a functional standpoint. At the system level, the system is represented as a whole rather than as in segments as in gate-level simulation. System-level simulation involves analyzing the entire system from a performance standpoint. It is this kind of simulation that we shall be concerned with in this book.

3.2 Why Simulation?

A large number of factors influence the decision to use any particular scientific technique to solve a given problem. The appropriateness of the technique is one consideration; economy is another. In this section, we consider the various advantages of using simulation as a modeling technique.

A system can be simplified to such an extent that it can be solved analytically. Such an analytical solution is desirable because it leads to a closed form solution (such as in Chapter 2), where the relationship between the variables is explicit. However, such a simplified form of the system is obtained by making several assumptions so as to make the solution mathematically tractable. Most real-life systems are so complex that some simplifying assumptions are not justifiable and we must resort to simulation. Simulation imitates the behavior of the system over time and provides data as if the real system were being observed.

Simulation as a modeling technique is attractive for the following reasons [4, 5]:

1. It is the next best thing to observing a real system in operation.
2. It enables the analysis of very complicated systems. A system can be so complex that its description by a mathematical model is beyond the capabilities of the analyst. "When all else fails" is a common slogan for many such simulations.
3. It is straight forward and easy to understand and apply. It does not rely heavily on mathematical abstractions that require an expert to understand and apply. It can be employed by many more individuals.
4. It is useful in experimenting with new or proposed designs prior to implementation. Once constructed, it may be used to analyze the

system under different conditions. Simulation can also be used in assessing and improving an existing system.

5. It is useful in verifying or reinforcing analytic solutions.

A major disadvantage of simulation is that it may be costly because it requires a large expenditure of time in construction, running, and validation. Another disadvantage is that simulation is difficult to validate and convince others. Because simulation is a slow and costly technique, it is sometimes referred to as the method of last resort. One should consider using simulation modeling when experimentation with the real system is expensive or dangerous, or when mathematical modeling of the system is intractable [2].

3.3 Characteristics of Simulation Models

As mentioned earlier, a model is a representation of a system. It can be a replica, a prototype, or a smaller-scale system [6]. For most analysis, it is not necessary to account for all different aspects of the system. A model simplifies the system to a sufficiently detailed level to permit valid conclusions to be drawn about the system. Depending on the objectives being pursued by the analyst, given system can be represented by several models. A wide variety of simulation models have been developed over the years for system analysis. To clarify the nature of these models, it is necessary to understand a number of characteristics.

3.3.1 Continuous/Discrete Models

This characteristic has to do with the model variables. A *continuous* model is one in which the state variables change continuously with time. The model is characterized by smooth changes in the system state. A *discrete* model is one in which state variables assume a discrete set of values. The model is characterized by discontinuous changes in the system state. The arrival process of messages in the queue of a LAN is discrete since the state variable, the number of waiting messages, changes only at the arrival or departure of a message.

3.3.2 Deterministic Stochastic Models

This characteristic deals with the system response. A system is *deterministic* if its response is completely determined by its initial state and input. It is *stochastic* (or non deterministic) if the system response may assume a range of values for given initial state and input. Thus, only the statistical averages of the output measures of a stochastic model are true characteristics of the real system. The simulation of a LAN usually involves random interarrival times and random service times.

3.3.3 Time/Event Based-Models

Since simulation is the dynamic portrayal of the states of a system over time, a simulation model must be driven by an automatic internal clock.

In *time-based* simulation, the simulation clock advances one "tick" of Δt. Figure 3.1 shows the flowchart of a typical time-based simulation model. Although time-based simulation is simple, it is inefficient because some action must take place at each clock "tick." An event signifies a change in the state of a system. In an *event-based* simulation model, updating only takes place at the occurrence of event, and the simulation clock is advanced by the amount of time since the last event. Thus, no two events can be processed at any pass. The need to determine which event is next in event-based simulation makes its programming complex. A problem with this type of simulation is that the speed at which the simulation proceeds is not directly related to real-time; correspondence to real-time operation is lost. Figure 3.2 is the flowchart of a typical event-based simulation.

The concepts of *event, process,* and *activity* are important in building a system model. As mentioned earlier, an event is an instantaneous occurrence that may change the state of the system. It may occur at an isolated point in time at which decisions are made to start or end an activity. A process is a time-ordered sequence of events. An activity represents a duration of time. The relationship of the three concepts is depicted in Figure 3.3 for a process that is comprised of five events and two activities. The concepts lead to three types of discrete simulation modeling [7, 8]: *event scheduling, activity scanning,* and *process interaction.*

3.3.4 Hardware/Software Models

Digital modeling may involve either hardware or software simulation. Hardware simulation involves using special-purpose equipment, with detailed programming reduced to a minimum. This equipment is sometimes called a *simulator.* In software simulation, the operation of the system is modeled using a computer program. The program describes certain aspects of the system that are of interest.

In this book, we are mainly concerned with software models that are discrete, stochastic, and event based.

3.4 Stages of Model Development

Once it has been decided that software simulation is the appropriate methodology to solve a particular problem, there are certain steps a model builder must take. These steps parallel the six stages involved in model development. (Note that the model is the computer program.) In programming terminology, these stages are [5, 9, 10] (1) model building, (2) program synthesis, (3) model verification, (4) model validation, (5) model analysis, and (6) documentation. The relationship of the stages is portrayed in Figure 3.4, where the numbers refer to the stages.

1. Model Building: This initial stage usually involves a thorough, detailed study of the system to decompose it into a manageable level of detail. The modeler often simplifies components or even omits some if their effects do not warrant inclusion. The task of the modeler is to produce a simplified yet valid abstraction of the system. This involves a careful study of the system of interest. The study should reveal interactions, dependence, and rules governing the components of the system. It should also reveal the estimation of the system variables and parameters. The modeler may use flowcharts to define or identify subsystems and their interactions. Since flowcharting is a helpful tool in describing a problem and planning a program, commonly used symbols are shown in Figure 3.5. These symbols are part of the flowcharting symbols formalized by the American National Standards Institute (ANSI). The modeler should feel free to adapt the symbols to his own style.

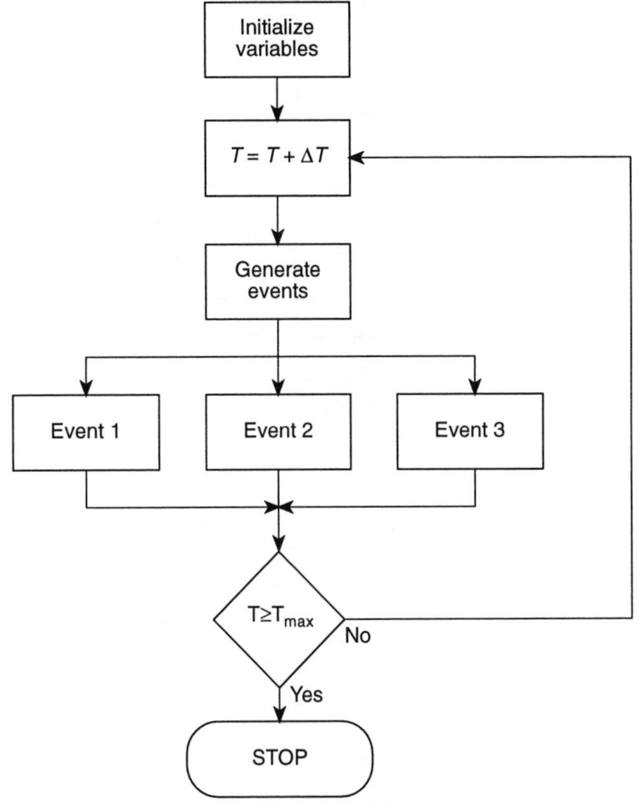

Figure 3.1 Typical time-based simulation model.

2. Program Synthesis: After a clear understanding of the simplified system and the interaction between components is gained, all the pieces are synthesized into a coherent description, which results in a computer program. The modeler must decide whether to implement the model in a general-purpose language such as PASCAL or FORTRAN or use a special-purpose simulation language such as GASP, GPSS, SLAM, SIMULA, SIMSCRIPT, or RESQ. A special-purpose simulation language usually requires less development time, but executes slower than does a general-purpose language. However, general-purpose languages so speed up programming and verification stages that they are becoming more and more popular in model development [5]. The type of computer and language used depend on the resources available to the programmer.

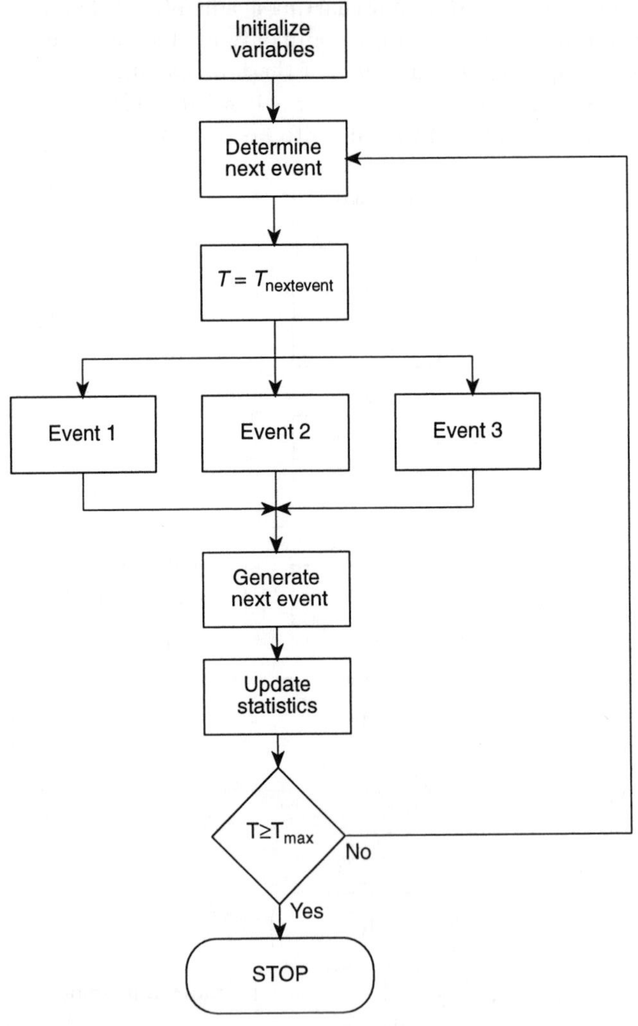

Figure 3.2 Typical event-based simulation model.

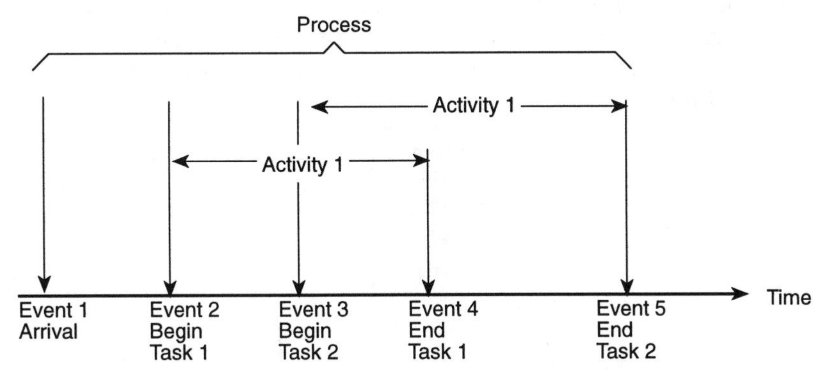

Figure 3.3 Relationship of events, activities, and processes.

3. Model Verification: This involves a logical proof of the correction of the program as a model. It entails debugging the simulation program and ensuring that the input parameters and logical structure of the model are correctly represented in the code. Although the programmer may know precisely what the program is intended to accomplish, the program may be doing something else.

4. Model Validation: This stage is the most crucial. Since models are simplified abstractions, their validity is important. A model is validated by proving that the model is a correct representation of the real system (verified program can represent an invalid model). This stage ensures that the computer model matches the real system by comparing the two. This is easy when the real system exists. It becomes difficult when the real system does not exist. In this case, a simulator can be used to predict the behavior of the real system. Validation may entail using a statistical proof of the correctness of the model. Whichever validation approach is used, validation must be performed before the model can be used. Validation may uncover further bugs and even necessitate reformulation of the model.

5. Model Analysis: Once the model has been validated, it can be applied to solve the problem at hand. This stage is the reason for constructing the model in the first place. It involves applying alternate input parameters to the program and observing their effects on the output parameters. The analysis provides estimate measures of the performance of the system.

6. Documentation: The results of the analysis must be clearly and concisely documented for future reference by the modeler or others. An inadequately documented program is usually useless to everyone including the modeler himself. The importance of this step cannot be overemphasized. A well-documented program or final report should be easily understood from the point of view of the simulation model. It is part of good documentation to insert comments here and there in the program, include sample input and output data, and list the assumptions and limitations of the model.

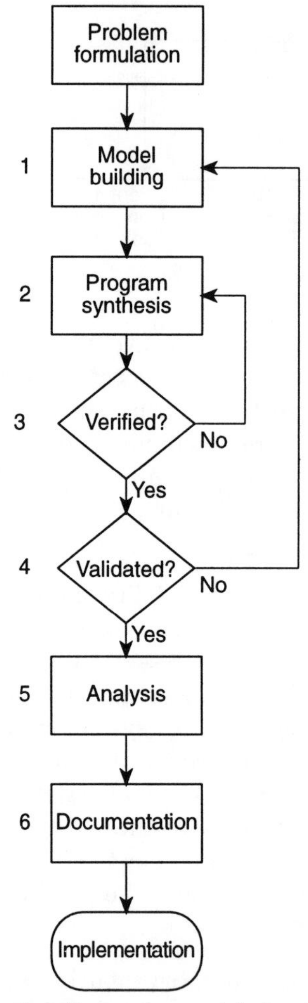

Figure 3.4 Stages in model development.

3.5 Common Mistakes in Simulation

Although it is impossible to enumerate all the likely mistakes one may make in developing a simulation model, it is important that we keep the common ones in mind. There are four major mistakes that may lead to erroneous results in a simulation model [11]:

- *Lack of understanding of the problem:* This is perhaps the most common mistake made by inexperienced analysts. It is usual for them to rush into developing a model without a thorough understanding of what they are simulating in the first place. One should resist this temptation. Before embarking on developing a model or writing the computer program for the model, one should try to understand all relevant theoretical background. The time spent in understanding the system should not be seen as wasted but part of the process. As the saying goes, "A problem well stated is half solved." A computer is no more than a tool used in the analysis of a program. For this reason, one should be as clear as possible about what one is really trying to model or what the computer is being asked to do before setting it off on several hours of expensive computations.
- *Ignoring important parameters:* The final outcome and validity of a model depend largely on the choices of the parameters describing the system one is modeling. Ignoring or overlooking important parameters may render the results unacceptable and the model a wasted effort. On the one hand, one must avoid incorporating all parameters relevant to the system, for this may result in a complex model that cannot be solved. A model that is simple and easy to explain is usually preferred to a complex one. On the other hand, one should avoid oversimplification of the system, which will make the results unacceptable.
- *Lack of adequate validation of the model:* It has been well said by A. C. Doyle that "It is a capital mistake to theoretize before you have all the evidence. It biases the judgement." Therefore, one should never trust the results of a simulation unless they are validated, at least in part. The results may be validated by comparing them with those obtained by previous investigators or with similar results obtained using a different approach, which may be analytical or numerical.
- *Poor documentation:* A poorly documented program or report does not benefit anyone other than the analyst himself. Both the program and final report on a simulation model should be properly documented in order to enable users other than the analyst to follow the logic of the model and the program. One common mistake is omitting assumptions and limitations in the documentation. This may cause the analyst or someone else to apply the model in a context where the assumptions are not valid.

3.6 Summary

This chapter has presented the basic concepts and definitions of simulation modeling of a system. The emphasis of the rest of the book is on discrete, stochastic, digital, software simulation modeling. It is discrete because it proceeds a step at a time. It is stochastic, or nondeterministic, because the element of randomness is introduced by using random numbers. It is digital because the computers employed are digital. It is a software simulation because the simulation model is a computer program.

Because simulation is a system approach to solving problems, we have considered the major stages involved in developing a model of a given system. These stages are model building, programming, verification, validation, analysis, and documentation.

Symbol	Meaning	
	Processing:	a group of operations; computation
	Decision:	a branching operation, usually an IF statement in FORTRAN
	Terminal:	marks the beginning or end of the program
	Connector:	an entry from, or point to, some other section of the flowchart

Figure 3.5 Common flowchart symbols.

References

[1] J. M. Carroll, *Simulation Using Personal Computers.* Englewood Cliffs, NJ: Prentice-Hall, 1987, p.2.

[2] F. Neelamkavil, *Computer Simulation and Modelling.* Chichester, UK: John Wiley & Sons, 1987, pp. 1–4, 12.

[3] M. H. MacDougall, *Simulating Computer Systems: Techniques and Tools.* Cambridge, MA: MIT Press, 1987, p.1.

[4] J. W. Schmidt and R. E. Taylor, *Simulation and Analysis of Industrial Systems.* Homewood, IL: R. D. Irwin, 1970, p. 5.

[5] J. Banks and J. S. Carson, *Discrete-Event System Simulation.* Englewood Cliffs, NJ: Prentice-Hall, 1984, pp. 3–16.

[6] W. Delaney and E. Vaccari, *Dynamic Models and Discrete Event Simulation.* New York, Marcel Dekker, 1989, pp. 1, 13.

[7] G. S. Fishman, *Concepts and Methods in Discrete Event Digital Simulation.* New York: John Wiley & Sons, 1973, pp. 24–25.

[8] A. A. B. Pritsker, *Introduction to Simulation and SLAM II.* New York: Halsted Press & Systems Publishing, 1984, 2nd ed., pp. 64–69.

[9] W. J. Graybeal and U. W. Pooch, *Simulation: Principles and Methods.* Cambridge, MA: Winthrop Publishers, 1980, pp. 5–10.

[10] T. G. Lewis and B. J. Smith, *Computer Principles of Modeling and Simulation.* Boston, MA: Houghton Mifflin, 1979, pp. 172–174.

[11] R. Jain, *The Art of Computer Systems Performance Analysis.* New York: John Wiley & Sons, 1991, pp. 14–22.

Problems

3.1 When should one use simulation as a method of solution?

3.2 What are the limitations of simulation models?

3.3 Define the following:
  (a) Discrete models
  (b) Stochastic models
  (c) Event-based models

3.4 Select a system and develop a simulation model following the stages discussed in Section 3.4. Justify your choices.

3.5 Produce a complete documentation for the computer program in Appendix A.

3.6 How would you validate the computer program in Appendix C ?

3.7 How would you debug your simulation model?

3.8 What are the differences between validation and verification of a simulation model? Which is more important?

3.9 Mention the major mistakes one makes in simulation modeling. How can they be avoided?

Chapter 4

Probability and Statistics

People can be placed in three classes: the few who make things happen; the many who watch things happen; and the overwhelming majority who have no idea of what has happened. We need more people who make things happen.

—Nicholas M. Butler

4.1 Introduction

This chapter provides the mathematical tools commonly used in simulation. As mentioned in the previous chapter, simulation is a procedure for analyzing systems with stochastic variables. In such situations, where randomness is the key factor, the outputs from simulation models are probabilistic and therefore require statistical estimates. This chapter briefly reviews important concepts in probability and statistics relating to simulation modeling, including

- Probability functions
- Random numbers generation
- Random variate generation from distribution functions
- Error estimation

It is perhaps necessary to summarize at the outset the following variables, which will be used throughout the chapter:

$$X = \text{the random variable}$$

$$P[\text{event } X] = \text{the probability that [event } X\text{] occurs}$$

$$p(x_i) \text{ or } f(x) = \text{the probability density function}$$

$$F(x) = \text{the cumulative density function}$$

$$E(X) = \mu = \text{the mean value of } X$$
$$V(X) = \sigma^2 = E[(X-\mu)^2] = \text{the variance of } X$$
$$\sigma = \text{the standard deviation of } X$$
$$\epsilon = \text{error estimate.}$$

4.2 Characterization of Random Variables

A random variable is defined as a function that assigns a real number to each outcome in the sample space. The random variable X is characterized by its *probability density function* (pdf) $f(x)$ or its *cumulative probability function* (cdf) $F(x)$, also known as its *probability distribution function*. A probability distribution function of a random variable is any rule that assigns a probability to each possible value of the variable. Thus, $F(x)$ is the probability measure for the event that consists of all possible outcomes with a value $X \leq x$; that is,

$$F(x) = P[X \leq x] \tag{4.1}$$

The properties of a cumulative distribution function $F(x)$ of a random variable X can be summarized as follows:

1. $F(x) \geq 0, \quad -\infty < x < \infty$
2. $F(-\infty) = 0$
3. $F(\infty) = 1$
4. $F(b) - F(a) = P[a < X \leq b] \geq 0, \quad \text{if } a < b$
5. $F(a) \leq F(b) \quad \text{if } a < b$

Note that these properties follow from the definition of $F(x)$ in (4.1). The first three properties are due the fact that $F(x)$ is a probability function. The last property shows that $F(x)$ is nondecreasing.

Whenever it exists, the derivative of $F(x)$ is the probability density function $f(x)$; that is,

$$f(x) = \frac{dF(x)}{dx} \tag{4.2}$$

or inversely

$$F(x) = \int_{-\infty}^{x} f(y)dy = P[X \leq x] \tag{4.3}$$

In differential notation

$$f(x)dx = dF(x) = P[x < X \leq x + dx] \tag{4.4}$$

The properties of the probability density function $f(x)$ of a random variable X can be readily derived from those for $F(x)$ and are summarized as follows:

1. $f(x) \geq 0$, since $F(x)$ is nondecreasing.
2. $\int_{-\infty}^{\infty} f(x)dx = 1$, since $F(\infty) = 1$.
3. $P[X \leq a] = F(a) = \int_{-\infty}^{a} f(x)dx$
4. $P[a < X \leq b] = F(b) - F(a) = \int_{a}^{b} f(x)dx$

The stochastic nature of a random variable is completely characterized by its cdf or pdf. It may also be characterized by the averages or *moments* of the random variable. The *first* moment of a random variable is its average value. It a measure of the "center of mass" of the probability. Thus the mean or expected value of a random variable X is defined as

$$\mu = E[X] = \begin{cases} \sum_i x_i p(x_i) & \text{in the discrete case} \\ \int_{-\infty}^{\infty} x f(x)dx & \text{in the continuous case} \end{cases} \tag{4.5}$$

It should be noted that μ is not the value one "expects" as the outcome of an experiment. Rather, it is the average value of the outcomes of several experiments. Some useful properties of the mean or expected value of X follow:

1. $E(aX + b) = E(aX) + E(b) = aE(X) + b$, if a and b are constants.
2. $E(X + Y) = E(X) + E(Y)$
3. $E(XY) = E[X]E[Y]$, if X and Y are independent variables.
4. $E[(X - \mu)^n] = n^{\text{th}}$ *central moment* of random variable X.

From the last property, we notice that the expected values of X^2 and X^3 are the *second* and *third* moments of X respectively.

Another important measure of the stochastic variability of X is its *variance* or *second central moment.* The variance of a random variable is the weighted average of the values of $(x - \mu)^2$. In other words, it is a measure of the spread of the random variable's pdf about its mean. It is mathematically defined as

$$\sigma^2 = V(X) = E[(X - \mu)^2] = \begin{cases} \sum_i (x_i - \mu)^2 p(x_i) & \text{discrete case} \\ \int_{-\infty}^{\infty} (x - \mu)^2 f(x)dx & \text{continuous case} \end{cases} \tag{4.6}$$

or

$$\begin{aligned} \sigma^2 &= E[(X - \mu)^2] \\ &= E(X^2 - 2\mu X + \mu^2) \\ &= E(X^2) - \mu^2 \end{aligned} \tag{4.7}$$

Note the following properties of variances:

1. $V(aX) = a^2 V(X)$, where a is a constant.
2. $V(X + a) = V(X)$
3. $V(X + Y) = V(X) + V(Y)$, if X and Y are independent random variables.

4. $V(aX + bY) = a^2V(X) + b^2V(Y)$, if X and Y are independent random variables.

The square root of the variance of a random number is the standard deviation σ of the random variable X. Hence,

$$\sigma = \sqrt{E[(X - \mu)^2]} \tag{4.8}$$

The variance and standard deviation measure the variability or dispersion of the random variable about its mean value. A low variance or standard deviation suggests that values of X are concentrated near the mean, whereas a high variance or standard deviation shows that the values are more spread out.

Example 4.1

Let X have the pdf

$$f(x) = \begin{cases} x & 0 < x < 1 \\ 2 - x & 1 < x < 2 \\ 0 & x > 2 \end{cases}$$

Find: (a) $F(x)$, (b) $P(\frac{1}{2} < X < \frac{3}{2})$, (c) $E(X)$.

Solution

(a) This is the triangular pdf (see Fig. 4.2). By definition,

$$F(x) = \int_{-\infty}^{x} f(y)dy = P(X \le x)$$

Hence $F(x) = 0, x \le 0$.

$$F(x) = \int_0^x ydy = \frac{x^2}{2}, \quad 0 \le x \le 1$$

$$F(x) = \int_0^1 ydy + \int_1^x (2 - y)dy$$
$$= 2x - \frac{x^2}{2} - 1, \quad 1 < x < 2$$

$$F(x) = 1, \quad x \ge 2.$$

(b) Also, by definition,

$$P(\frac{1}{2} < X < \frac{3}{2}) = F(\frac{3}{2}) - F(\frac{1}{2})$$
$$= 0.875 - 0.125 = 0.75$$

$$(c) \quad E(X) = \int_{-\infty}^{\infty} xf(x)dx = \int_0^1 x^2dx + \int_1^2 x(2-x)dx$$
$$= \frac{1}{3} + \frac{2}{3} = 1$$

4.3 Common Probability Distributions

The preceeding section has discussed probability distribution functions in general. Here we consider seven specific distributions that are commonly used in simulation.

4.3.1 Uniform Distribution

This distribution, also known as the rectangular distribution, is one in which the density is constant. It models random events in which every value between a minimum and maximum value is equally likely. Its graph is in Figure 4.1. Its characteristics are as follows:

$$\begin{aligned} f(x) &= \frac{1}{b-a}, \quad a \le x \le b \\ F(x) &= \frac{x-a}{b-a}, \quad a \le x \le b \\ E(X) &= \frac{b+a}{2} \\ V(X) &= \frac{(b-a)^2}{12} \end{aligned} \tag{4.9}$$

A special uniform distribution for which $a = 0$ and $b = 1$, called the standard uniform distribution, is very useful in generating random samples from any probability distribution function.

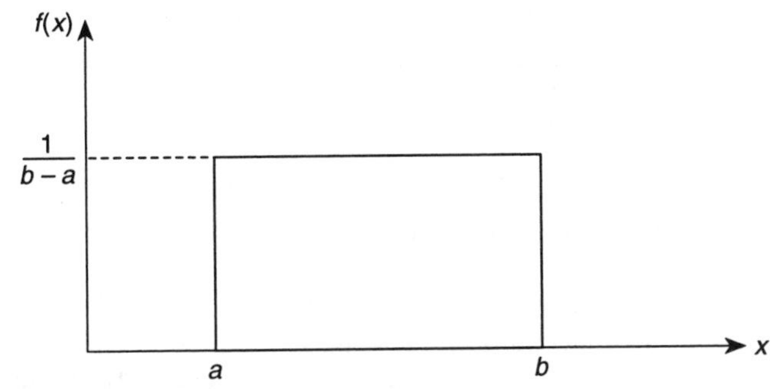

Figure 4.1 Uniform density function.

4.3.2 Triangular Distribution

This distribution is specified by three parameters: a minimum, a mode, and a maximum. The mode, which occurs at $x = b$, is often used to characterize the distribution. The density function for the triangular distribution is shown in Figure 4.2. The distribution has the following properties:

$$
\begin{aligned}
f(x) &= \begin{cases} \dfrac{2(x-a)}{(b-a)(c-a)}, & a \le x \le b \\ \dfrac{2(c-x)}{(c-b)(c-a)}, & b \le x \le c \end{cases} \\
F(x) &= \begin{cases} \dfrac{(x-a)^2}{(b-a)(c-a)}, & a \le x \le b \\ 1 - \dfrac{(c-x)^2}{(c-b)(c-a)}, & b \le x \le c \end{cases} \\
E(X) &= \frac{a+b+c}{3} \\
V(X) &= \frac{(a^2+b^2+c^2) - ab - ac - bc}{18}
\end{aligned}
\tag{4.10}
$$

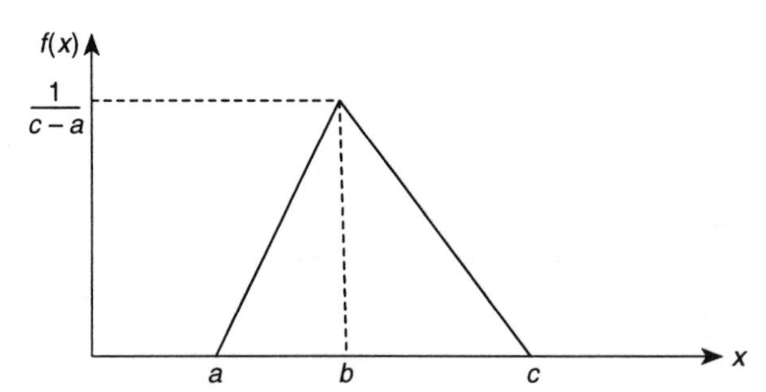

Figure 4.2 Triangular density function.

4.3.3 Exponential Distribution

This distribution, also known as the negative exponential distribution, is frequently used in simulations of queueing systems to describe the interarrival or interdeparture times of customers at a server. Its frequent use is because the remaining time is not conditioned on expended time. This peculiar characteristic is known variably as the Markov (*forgetfulness* or *lack of memory*) property. The distribution is portrayed in Figure 4.3

and has the following characteristics:

$$\begin{aligned} f(x) &= \alpha e^{-\alpha x}, \quad x > 0,\ \alpha > 0 \\ F(x) &= 1 - e^{-\alpha x}, \quad x > 0 \\ E(X) &= \frac{1}{\alpha} \\ V(X) &= \frac{1}{\alpha^2} \end{aligned} \tag{4.11}$$

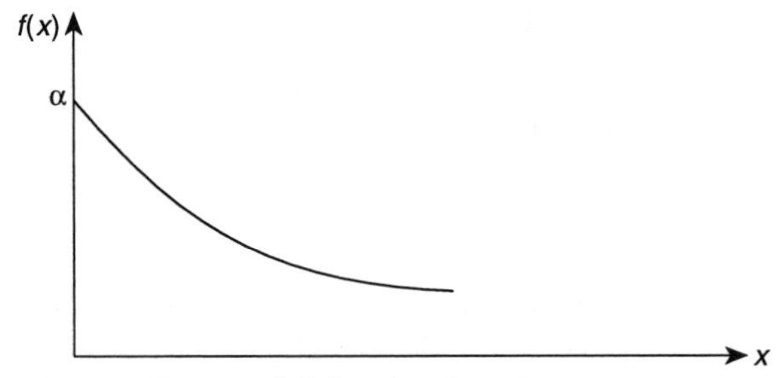

Figure 4.3 Exponetial density function.

4.3.4 Poisson Distribution

This is a discrete distribution widely used in modeling arrival distributions and other random events. It is particularly important in the analysis of queueing systems. Its pdf denotes the probability that event X occurs in an interval of fixed length. If γ is the mean number of occurences in the interval under consideration, then

$$\begin{aligned} P_n(\gamma) &= \frac{\gamma^n e^{-\gamma}}{n!}, \quad n = 0, 1, 2, \ldots \\ E(X) &= \gamma \\ V(X) &= \gamma \end{aligned} \tag{4.12}$$

Notice that only the Poisson distribution has equal mean and variance. Figure 4.4 shows the Poisson distribution for $\gamma = 3$. As mentioned before, the Poisson distribution is used to express the probability of finding n occurences in an interval of fixed length. To incorporate the interval itself, the Poisson distribution is usually written as

$$P_n(t) = \frac{(\lambda t)^n e^{-\lambda t}}{n!}, \quad n = 0, 1, 2, \ldots \tag{4.13}$$

If t denotes time, $P_n(t)$ is the probability of n successes in time t and λ is the success rate or event frequency, that is, the mean number of events per unit time.

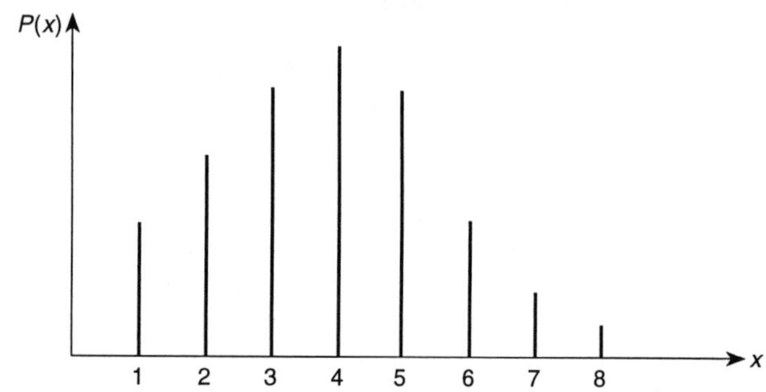

Figure 4.4 Poisson distribution for fixed γ.

The Poisson distribution is related to other probalility distribution as follows:

1. It is the limiting case of the Bernoulli probability function under the following conditions: The number of trials $m \to \infty$, the probability of success in one trial $p \to 0$ (meaning that success is a rare event), and the expected number of successes is a finite number $\gamma = mp$.
2. For a large mean, the Poisson distribution is approximated by the normal distribution.
3. It is related to the exponential distribution. If the time interval between outcomes is exponentially distributed, the number that occurs in a fixed time interval is Poisson distributed. In other words, if interarrival distribution is exponential, the distribution of the number of arrivals will be Poisson. The converse is also true. If the number of arrivals (or number of points occurring) in a time interval t is given by the Poisson distribution in Eq. (4.12) with mean λ, then the interarrival times (or the distances between successive points) are exponential variables with mean $1/\lambda$.

4.3.5 Normal Distribution

This distribution, also known as the Gaussian distribution, is heavily used in probability and statistics, for reasons that will soon become obvious. It is used to describe phenomena with symmetric variations above and below the mean μ. The normal distribution has the form

$$f(x) = \frac{1}{\sigma\sqrt{2\pi}} \exp\Big[-\frac{1}{2}\Big(\frac{x-\mu}{\sigma}\Big)^2\Big], \quad -\infty < x < \infty \tag{4.14}$$

where the mean μ and the variance σ^2 are themselves incorporated in the pdf. Figure 4.5 shows the normal pdf. It is a common practice to use the notation $X \sim N(\mu, \sigma^2)$ to denote a normal random variable X with mean μ and variance σ^2. When $\mu = 1$ and $\sigma = 0$, we have the *standard normal* distribution function with

$$f(x) = \frac{1}{\sqrt{2\pi}} e^{-x^2/2} \tag{4.15}$$

which is widely tabulated [1–3].

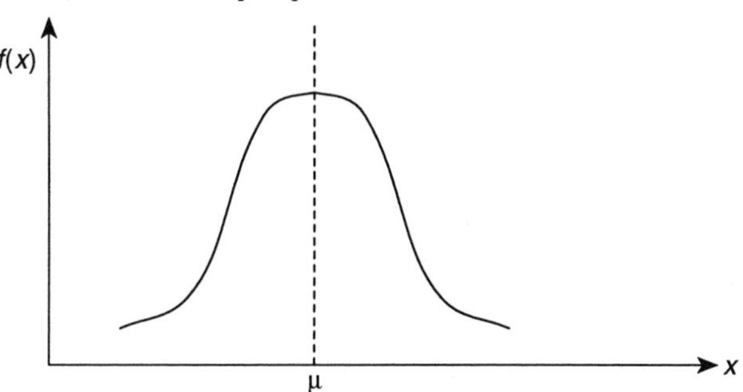

Figure 4.5 Normal or Gaussian density function.

It is important that we note the following points about the normal distribution, which make the distribution the most prominent in probability and statistics.

1. The binomial probability function with parameters m and p is approximated by a Gaussian pdf with $\mu = mp$ and $\sigma^2 = mp(1-p)$ for large m and finite p.
2. As mentioned earlier, the Poisson probability function with parameter γ can be approximated by a normal distribution with $\mu = \sigma^2 = \gamma$ for large γ.
3. The normal distribution is useful in characterizing the uncertainty associated with the estimated values. In other words, it is used in performing statistical analysis on simulation output.
4. The justification for the use of normal distribution comes from the *central limit theorem,* which, as will be discussed in Section 4.6, states that the distribution of the sum of n independent variables from any distribution approaches a normal distribution as n becomes large.

Thus, the normal distribution is used to model the cumulative effect of many small disburbances, each of which contributes to the stochastic variable X. It has the advantage of being mathematically tractable. Consequently, many statistical analyses such as those of regression and variance have been derived assuming a normal density function.

4.3.6 Lognormal Distribution

This is a distribution of a variable whose natural logarithm follows the normal distribution. In other words, if a random variable x follows a lognormal distribution with parameters μ and σ, then $\ln x$ is normally distributed with variable $N(\mu, \sigma)$. The lognormal distribution has the form

$$f(x) = \frac{1}{\sigma x \sqrt{2\pi}} \exp\left[-\frac{1}{2}(\frac{\ln x - \mu}{\sigma})^2\right], \quad -\infty < x < \infty$$

with

$$E(X) = \exp[\mu + \sigma^2/2]$$
$$V(X) = \exp[2\mu + \sigma^2](e^{\sigma^2} - 1) \tag{4.16}$$

Figure 4.6 illustrates a lognormal distribution. Using the central limit theorem, it can be shown that the distribution of the product of independent random variables approaches a lognormal distribution. Just as the normal distribution models an additive process, the lognormal distributiom models a multiplicative process.

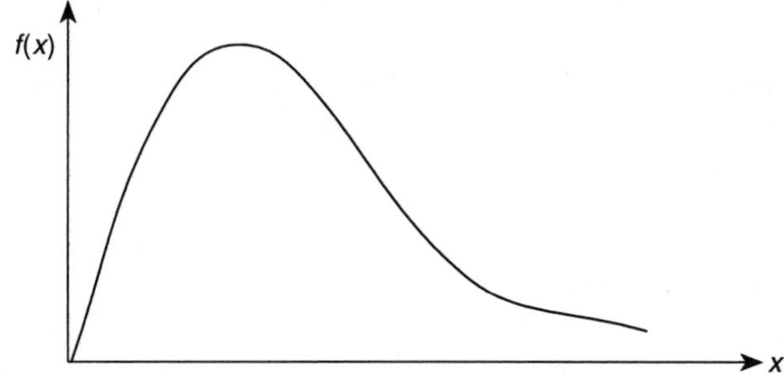

Figure 4.6 Lognormal density function.

4.3.7 Gamma (Erlang) Distribution

The Erlang distribution is widely used in queueing theory to model an activity that occurs in phases, with each phase being exponentially distributed. It is derived as the sum (or convolution) of independent, identically distributed exponential random variables. The gamma distribution is a generalization of the Erlang distribution where the number of sums of exponentials need not be an integer. The distribution has the following properties:

$$f(x) = \frac{\alpha^k x^{k-1} e^{-\alpha x}}{\Gamma(k)}, \quad x > 0,\ \mu > 0,\ \alpha > 0$$
$$E(X) = k/\alpha$$
$$V(X) = k/\alpha^2 \tag{4.16}$$

where α and k are called the *shape parameters.* The function $\Gamma(k)$ is the gamma function, and k may or may not be an integer. If k is an integer,

$\Gamma(k) = (k-1)!$ Figure 4.7 show the gamma density function for positive, negative, and zero values of k. By using different shape parameters, the gamma distribution can be made to approach a variety of distributions and consequently can represent different physical processes. Notice that

1. When $k = 1$, the gamma distribution reduces to the exponential distribution.
2. When k is a positive integer, the gamma distribution becomes the Erlang distribution.
3. The sum of squares of normal random variables, which is the chi-squared distribution, is a special case of the gamma distribution.

Thus, we observe that the special cases of the gamma distribution are the Erlang, exponential, and chi-squared distributions.

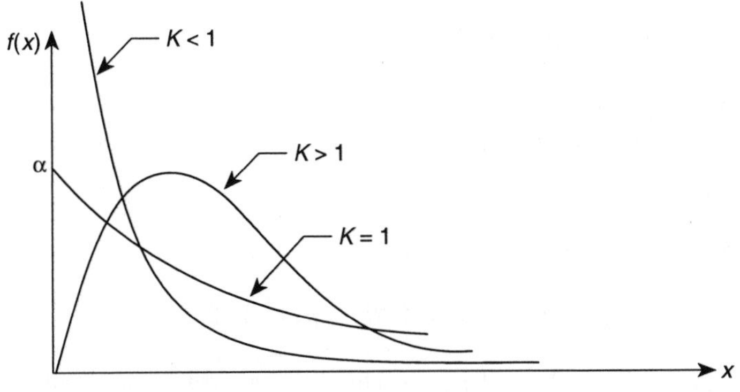

Figure 4.7 Gamma density Function.

Example 4.2

Customers arrive at a bank at the rate of $\frac{1}{3}$ customer per minute. If X denotes the number of customers to arrive in the next 9 minutes, calculate the probability that (a) there will be no customers within that period, (b) exactly 3 customers will arrive in this period, and (c) at least 4 customers will arrive.

Solution

This is a Poisson process with parameter $\gamma = \lambda t = \frac{1}{3}(9) = 3$ and its probability function is

$$P_n(\gamma) = \frac{3^n e^{-3}}{n!}$$

(a)

$$P_0(\gamma) = \frac{3^0 e^{-3}}{0!} = 0.05$$

(b)

$$P_3(\gamma) = \frac{3^3 e^{-3}}{3!} = 0.224$$

(c) The probability that at least 4 customers arrive is

$$\sum_{n=4}^{\infty} \frac{3^n e^{-3}}{n!} = 1 - P_0 - P_1 - P_2 - P_3 = 0.353$$

Example 4.3

For messages that are geometrically distributed, the probability that a message is n units in length is

$$P(X = n) = pq^{n-1} \quad q = 1 - p, \; n = 1, 2, 3, \ldots$$

(a) Sketch the discrete probability distribution for $p = q = 0.5$.
(b) Find $E(X)$ and $V(X)$.

Solution

(a) If we let $p_n = P(X = n)$, for $p = q = 0.5$,

$$p_1 = 0.5, \quad p_2 = 0.25, \quad p_3 = 0.125, \ldots$$

that is, p_n is always half of p_{n-1}. Thus, the p_n are as sketched in Figure 4.8. Notice that the probabilities decrease in geometric proportion, hence the name goemetric distribution.
(b) By definition,

$$E(X) = \sum_{n=1}^{\infty} nP(X = n) = \sum_{n=1}^{\infty} npq^{n-1} = p \sum_{n=1}^{\infty} nq^{n-1}$$

Recall that

$$\sum_{n=1}^{\infty} a^n = \frac{a}{1-a}$$

Taking derivatives of both sides with respect to a gives

$$\sum_{n=1}^{\infty} na^{n-1} = \frac{1}{(1-a)^2}$$

Applying this gives

$$E(X) = \frac{p}{(1-q)^2} = \frac{1}{p}$$

Similarly,

$$E[X(X-1)] = \sum_{n=1}^{\infty} n(n-1)pq^{n-1} = \frac{2pq}{(1-q)^3} = \frac{2q}{p^2}$$

so that

$$E(X^2) = E[X(X-1)] + E(X) = \frac{2q}{p^2} + \frac{1}{p}$$

and thus

$$V(X) = E(X^2) - E(X)^2 = \frac{2q}{p^2} + \frac{1}{p} - \frac{1}{p^2} = \frac{q}{p^2}$$

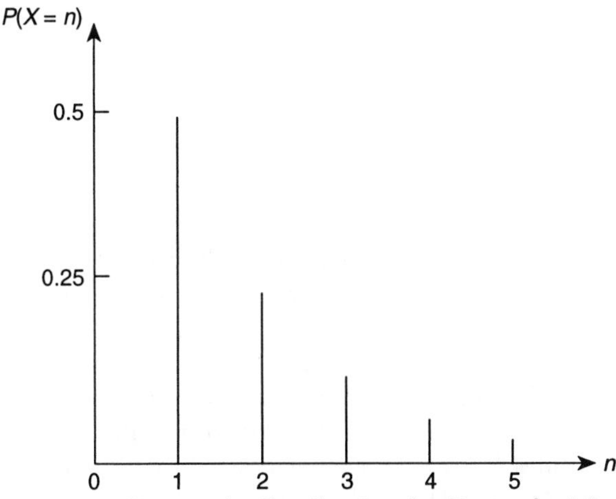

Figure 4.8 Geometric distribution for Example 4.3

4.4 Generation of Random Numbers

Due to the stochastic nature of a simulation model, the simulation must include a mechanism for generating random *variates*—variables whose values obey a specified probability distribution. The generation of random variates requires the availability of random numbers. Hence this section deals with procedures for generating random numbers, and the next section deals with the process of translating the random numbers into random variates.

There are various techniques by which random numbers have been and are being generated [4–11]. The almost universally used method of generating random numbers is to select a function $H(Z)$ that maps integers into random numbers. Select some guessed value Z_0, and generate the next random number as $Z_{k+1} = H(Z_k)$. The commonest function $H(Z)$ takes the form

$$H(Z) = (aZ + c) \bmod m \tag{4.18}$$

where

$$\begin{aligned} Z_0 &= \text{a starting value or a seed } (Z_0 > 0) \\ a &= \text{ multiplier } (a \geq 0) \\ c &= \text{increment } (c \geq 0) \\ m &= \text{the modulus} \end{aligned}$$

The modulus m is usually 2^t for t-digit binary integers. For a 31-bit computer machine, for example, m may be 2^{31-1}. Here Z_0, a, and c are integers in the same range, as $m > a$, $m > c$, $m > Z_0$. The desired sequence of random numbers Z_n is obtained from

$$Z_{n+1} = (aZ_n + c) \bmod m \tag{4.19}$$

This is called a *linear congruential sequence.* For example, if $Z_0 = a = c = 7$ and $m = 10$, the sequence is

$$7,\ 6,\ 9,\ 0,\ 7,\ 6,\ 9,\ 0,\ \ldots \tag{4.20}$$

It is evident that congruential sequences always get into a loop; that is, there is ultimately a cycle of numbers which is repeated endlessly. The sequence in Eq. (4.20) has a period of length 4. A useful sequence will of course have a relatively long period. The terms *multiplicative congruential method* and *mixed congruential method* are used by many authors to denote linear congruential methods with $c = 0$ and $c \neq 0$ respectively. Rules for selecting $Z_0, a, c,$ and m can be found in Knuth [6] and Law and Kelton [10].

Here we are interested in generating random numbers from the uniform distribution in the interval $(0, 1)$. These numbers will be designated by the letter U and are obtained from Eq. (4.19) as

$$U = \frac{Z_{n+1}}{m} \tag{4.21}$$

Thus U can only assume values from the set $\{0, 1/m, 2/m, ..., (m-1)/m\}$. For generating random numbers X uniformly distributed in the interval (a, b), we use

$$X = a + (b - a)\ U \tag{4.22}$$

Random numbers that are based on mathematical relations in Eqs. (4.19) and (4.21) and produced by computer are not truly random; in fact, given the seed of the sequence, all numbers U of the sequence are completely predictable or deterministic. Some authors emphasize this point by calling such computer generated sequences *pseudorandom numbers.* However, with a good choice of $a, c,$ and m, the sequences of U appear to be sufficiently random in that they pass a series of statistical tests of randomness. They have the advantage over truly random numbers of being generated in a fast way and of being reproducible, when desired, especially for program debugging.

Pseudorandom numbers that are adequate for experimental purposes must satisfy some criteria. A good generator must be capable of producing random numbers that have any of m discrete values, which are equally likely to occur (i.e., the probability of a value's occurence is $1/m$), and each new value is completely independent of any previous output of the generator. Statistically, this implies that the numbers are *uniformly* distributed, *independent* variables. It is also significant that the random numbers contain enough digits so that generation of numbers on the interval $\{a, b\}$ is sufficiently *dense*. A battery of statistical tests has been developed to reveal departures from independence and uniformity or to determine whether a series of numbers meets the criterion of randomness. For random numbers in the interval (0, 1), a quick and simple test of the randomness is that the mean is 0.5. Other tests can be found in Fishman [4], Knuth [6], Banks and Carson [7] and Neelamkavil [11].

In most cases, a modeler need not generate uniform random numbers. Random number generators are usually built into a simulation package or available in a digital computer.

Example 4.4

(a) Using a linear congruential scheme, generate 10 pseudorandom numbers with $a = 1,573$, $c = 19$, $m = 10^3$, and seed value $X_0 = 89$. (b) Repeat the generation with $c = 0$.

Solution

(a) Substituting $a = 1,573$, $c = 19$, $m = 1{,}000$, and $X_0 = 89$ in Eq. (4.18) leads to

$$X_1 = 1,573 \times 89 + 19 \text{ (mod 1,000)} = 16$$
$$X_2 = 1,573 \times 16 + 19 \text{ (mod 1,000)} = 187$$
$$X_3 = 1,573 \times 187 + 19 \text{ (mod 1,000)} = 170$$
$$X_4 = 1,573 \times 170 + 19 \text{ (mod 1,000)} = 429$$
$$X_5 = 1,573 \times 429 + 19 \text{ (mod 1,000)} = 836$$
$$X_6 = 1,573 \times 836 + 19 \text{ (mod 1,000)} = 47$$
$$X_7 = 1,573 \times 47 + 19 \text{ (mod 1,000)} = 950$$
$$X_8 = 1,573 \times 950 + 19 \text{ (mod 1,000)} = 369$$
$$X_9 = 1,573 \times 369 + 19 \text{ (mod 1,000)} = 456$$
$$X_{10} = 1,573 \times 456 + 19 \text{ (mod 1,000)} = 307$$

(b) For $c = 0$, we obtain

$$
\begin{aligned}
X_1 &= 1,573 \times 89 \text{ (mod 1,000)} = 997 \\
X_2 &= 1,573 \times 997 \text{ (mod 1,000)} = 281 \\
X_3 &= 1,573 \times 281 \text{ (mod 1,000)} = 13 \\
X_4 &= 1,573 \times 13 \text{ (mod 1,000)} = 449 \\
X_5 &= 1,573 \times 449 \text{ (mod 1,000)} = 277 \\
X_6 &= 1,573 \times 277 \text{ (mod 1,000)} = 721 \\
X_7 &= 1,573 \times 721 \text{ (mod 1,000)} = 133 \\
X_8 &= 1,573 \times 133 \text{ (mod 1,000)} = 209 \\
X_9 &= 1,573 \times 209 \text{ (mod 1,000)} = 757 \\
X_{10} &= 1,573 \times 759 \text{ (mod 1,000)} = 761
\end{aligned}
$$

4.5 Generation of Random Variates

Having discussed how to generate random numbers, we now discuss various procedures for converting the random numbers into random variates.

It is usually required in a simulation to generate a random variable X from a given probability distribution $F(x)$. This can be accomplished using several techniques [6, 12–16] such as the *inverse transformation*, the *rejection, rectangular approximation, convolution,* and *look-up* methods. The most commonly used techniques are the inverse transformation method and the rejection method.

4.5.1 Inverse Transformation Method

The inverse transformation method, otherwise known as the direct method, is the most straightforward technique for generating random variates from probability distribution functions. It basically entails inverting the cumulative probability function $F(x) = P[X \leq x]$ associated with the random variable X. The fact that $0 \leq F(x) \leq 1$ intuitively suggests that by generating random numbers U uniformly distributed over (0, 1) we can produce a random sample X from the distribution of $F(x)$ by inversion. Thus, to generate random X with probability distribution $F(x)$, we set $U = F(x)$ and obtain

$$X = F^{-1}(U) \tag{4.23}$$

where X has the distribution function $F(x)$. For example, if X is a random variable that is exponentially distributed with mean μ, then

$$F(x) = 1 - e^{-x/\mu}, \qquad 0 < x < \infty \tag{4.24}$$

Solving for X in $U = F(X)$ gives

$$X = -\mu \ \ln\left(1 - U\right) \tag{4.25}$$

Since $(1 - U)$ is itself a random number in the interval (0, 1), we simply write

$$X = -\mu \ \ln U \tag{4.26}$$

This technique for generating random variates assumes that the inverse transformation $F^{-1}(U)$, required in Eq. (4.23), exists. This makes it suitable for generating random variates from exponential, gamma, uniform, and empirical distributions. However, there are several distributions for which $F^{-1}(U)$ cannot be found analytically and the inverse transformation method cannot be used. A typical example is the normal distribution.

4.5.2 Rejection Method

This technique, also known as the rejection-acceptance method, can be applied to the probability distribution of any bounded variable. To apply the method, we let the probability density function of the random variable $f(x) = 0$ for $a > x > b$ and let $f(x)$ be bounded by M (i.e., $f(x) \leq M$) as shown in Figure 4.9. We generate random variates by taking the following steps:

1. Generate two random numbers (U_1, U_2) in the interval (0, 1).
2. Compute two random numbers with uniform distributions in (a, b) and $(0, M)$ respectively: that is
   $X = a + (b - a)U_1$ (scale the variable on the x-axis)
   $Y = U_2 M$ (scale the variable on the y-axis).
3. If $Y \leq f(X_1)$, accept X the next random variate, otherwise reject X and return to Step 1.

In the rejection technique all points falling above $f(x)$ are rejected, and those points falling on or below $f(x)$ are utilized to generate X through $X = a + (b - a)\ U_1$. The efficiency of the method is enhanced as the probability of acceptance in Step 3 increases.

How the inverse transformation method and the rejection methods are used in generating random variates for specific distribution functions is presented in Fishman [4], Banks and Carson [7], and Taha [16].

Example 4.5

The binomial distribution represents the probability that an event or successful outcome occurs n times out of m trials.

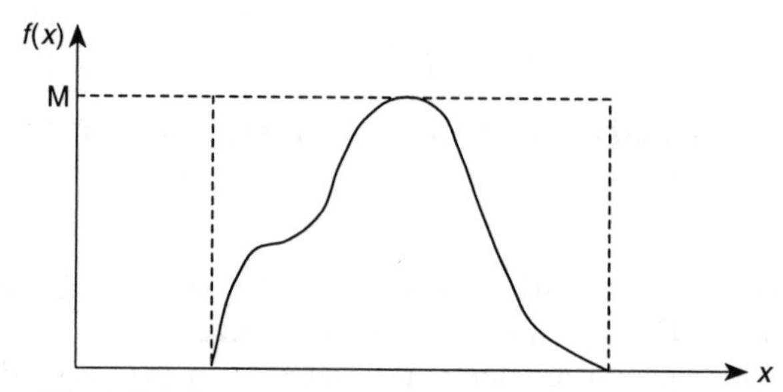

Figure 4.9 The rejection method of generating a random variate from $f(x)$.

$$p_n = \binom{m}{n} p^k q^{m-n}, \quad q = 1 - p, \quad n = 0, 1, 2, \ldots$$

where p is the probability of a successful outcome in each trial. Provide an algorithm for generating a binomial variate.

Solution

There are many ways of generating a binomial variate X. One way is described as follows [17]:

1. Given values of p and m, set $X_0 = 0$.
2. Generate a random number U $(0 < U < 1)$, and set

$$X_i = \begin{cases} X_{i-1} + 1, & \text{if} \quad U \leq p \\ X_{i-1}, & \text{if} \quad U > p \end{cases}$$

3. Repeat step 2 for n times and set $X_n = X$. X is the required binomial variate.

This method can be used to generate as many binomial variates as needed.

4.6 Estimation of Error

Simulation procedures give solutions which are averages over a number of tests. For this reason, it is important to realize that the sample statistics obtained from simulation experiments will vary from one experiment to another. In fact, the sample statistics themselves are random variables and, as such, have associated probability distributions, means, variances, and standard deviation. Thus the simulation results contain fluctuations about a mean value, and it is impossible to ascribe a 100% confidence in the results. To evaluate the statistical uncertainty or error in a simulation experiment, we must resort to various statistical techniques associated with random variables and utilize the central limit theorem.

Suppose that X is a random variable. You recall that we define the expected or mean value of X as

$$\mu = \int_{-\infty}^{\infty} x f(x) dx \tag{4.27}$$

where $f(x)$ is the probability density function of X. If we draw random and independent samples, $x_1, x_2, ..., x_N$ from $f(x)$, our estimate of x would take the form of the mean of N samples, namely,

$$\hat{\mu} = \frac{1}{N} \sum_{n=1}^{N} x_n \tag{4.28}$$

Whereas μ is the true mean value of X, $\hat{\mu}$ is the unbiased estimator of μ—an unbiased estimator being one with the correct expectation value. The expected value $\hat{\mu}$ is close to μ, but $\hat{\mu} \neq \mu$. The standard deviation, defined as

$$\sigma(x) = \left[E(X^2) - \mu^2\right]^{1/2} \tag{4.29}$$

provides a measure of the spread in the values of $\hat{\mu}$ about μ; it yields the order of magnitude of the error. The confidence we place in the estimate of the mean is given by the variance of $\hat{\mu}$. The relationship between the variance of $\hat{\mu}$ and the variance of x is

$$\sigma(\hat{\mu}) = \frac{\sigma(x)}{\sqrt{N}} \tag{4.30}$$

This shows that if we use $\hat{\mu}$ constructed from N values of x_n according to Eq. (4.28) to estimate μ, then the spread in our results of $\hat{\mu}$ about μ is proportional to $\sigma(x)$ and falls off as the number of samples N increases.

In order to estimate the spread in $\hat{\mu}$, we define the *sample variance*

$$S^2 = \frac{1}{N-1} \sum_{n=1}^{N} (x_n - \hat{\mu})^2 \tag{4.31}$$

Again, it can be shown that the expected value of S^2 is equal to $\sigma^2(x)$. Therefore the sample variance is an unbiased estimator of $\sigma^2(x)$. Multiplying out the square term in Eq. (4.31), it is readily shown that the *sample standard deviation* is

$$S = \left(\frac{N}{N-1}\right)^{1/2} \left[\frac{1}{N}\sum_{n=1}^{N} x_n^2 \; - \; \hat{x}^2\right]^{1/2} \tag{4.32}$$

For large N, the factor $N/(N-1)$ is set equal to one.

As a way of arriving at the central limit theorem, a fundamental result in probability theory, consider the binomial function

$$B(M) = \frac{N!}{M!(N-M)!} p^M q^{N-M} \tag{4.33}$$

which is the probability of M successes in N independent trials. In Eq. (4.33), p is the probability of success in a trial and $q = 1 - p$ is the probability of failure. If M and $N - M$ are large, we may use *Stirling's formula*

$$n! \sim n^n e^{-n} \sqrt{2\pi n} \tag{4.34}$$

so that Eq. (4.33) is approximated as the normal distribution [18]:

$$B(M) \simeq f(\hat{x}) = \frac{1}{\sigma(\hat{x})\sqrt{2\pi}} \exp\left[-\frac{(\hat{x} - \bar{x})^2}{2\sigma^2(\hat{x})}\right] \tag{4.35}$$

where $\bar{x} = Np$ and $\sigma(\hat{x}) = \sqrt{Npq}$. Thus, as $N \to \infty$, the central limit theorem states that the probability density function that describes the distribution of $\hat{x}$ that results from N simulation experiments is the normal distribution $f(\hat{x})$ in Eq. (4.35). In other words, the sum of a large number of random variables tends to be normally distributed. Inserting Eq. (4.30) into Eq. (4.35) gives

$$f(\hat{\mu}) = \sqrt{\frac{N}{2\pi}} \frac{1}{\sigma(x)} \exp\left[-\frac{N(\hat{\mu} - \mu)^2}{2\sigma^2(x)}\right] \tag{4.36}$$

The normal (or Gaussian) distribution is very useful in various problems in engineering, physics, and statistics. The remarkable versatility of the Gaussian model stems from the central limit theorem. For this reason, the Gaussian model often applies to situations in which the quantity of interest results from the summation of many irregular and fluctuating components.

Since the number of samples N is finite, absolute certainty in simulation is unattainable. We try to estimate some limit or interval around μ such that we can predict with some confidence that $\hat{\mu}$ falls within that limit. Suppose we want the probability that $\hat{\mu}$ lies between $\mu - \epsilon$ and $\mu + \epsilon$. By definition,

$$P\left[\mu - \epsilon < \hat{\mu} < \mu + \epsilon\right] = \int_{\mu-\epsilon}^{\mu+\epsilon} f(\hat{\mu}) d\hat{\mu}. \tag{4.37}$$

By letting $\lambda = \dfrac{(\hat{\mu} - \mu)}{\sqrt{2/N}\sigma(x)}$, we get

$$P[\mu - \epsilon < \hat{\mu} < \mu + \epsilon] = \frac{2}{\sqrt{\pi}} \int_0^{(\sqrt{N/2})(\epsilon/\sigma)} e^{-\lambda^2} d\lambda$$

$$= \operatorname{erf}\left(\sqrt{N/2}\frac{\epsilon}{\sigma(x)}\right) \tag{4.38}$$

or

$$P[\mu - z_{\alpha/2}\frac{\sigma}{\sqrt{N}} \leq \hat{\mu} \leq \mu + z_{\alpha/2}\frac{\sigma}{\sqrt{N}}] = 1 - \alpha \tag{4.39}$$

where erf(x) is the error function and $z_{(\alpha/2)}$ is the upper $\alpha/2 \times 100$ percentile of the standard normal deviation. The random interval $\hat{x} \pm \epsilon$ is called a *confidence interval* and erf $\left(\sqrt{N/2}\right)[\epsilon/\sigma(x)]$ is the *confidence level*.

Most simulation experiments use error $\epsilon = \sigma(x)/\sqrt{N}$ which implies that $\hat{\mu}$ is within one standard deviation of μ, the true mean. From Eq. (4.39), the probability that the sample mean $\hat{\mu}$ lies within the interval $\hat{\mu} \pm \sigma(x)/\sqrt{N}$ is 0.6826 or 68.3%. If higher confidence levels are desired, two or three standard deviations may be used. For example,

$$P\left[\mu - M\frac{\sigma(x)}{\sqrt{N}} < \hat{\mu} < \mu + M\frac{\sigma(x)}{\sqrt{N}}\right] = \begin{cases} 0.6826, & M = 1 \\ 0.954, & M = 2 \\ 0.997, & M = 3 \end{cases} \tag{4.40}$$

where M is the number of standard deviations.

In Eqs. (4.39) and (4.40), it is assumed that the population standard deviation σ is known. Since this is rarely the case, σ must be estimated by the sample standard S calculated from Eq. (4.32), so the normal distribution is replaced by the student's t-distribution. It is well known that the t-distribution approaches the normal distribution as N becomes large, say $N > 30$. Thus Eq. (4.39) is equivalent to

$$P[\mu - \frac{S\, t_{\alpha/2;N-1}}{\sqrt{N}} \leq \hat{\mu} \leq \mu + \frac{S\, t_{\alpha/2;N-1}}{\sqrt{N}}] = 1 - \alpha \tag{4.41}$$

where $t_{\alpha/2;N-1}$ is the upper $100\times(\alpha/2)$ percentage point of the student's t-distribution with $(N - 1)$ degrees of freedom. Its values are listed in any standard statistics text.

The confidence interval $\hat{x} - \epsilon < x < \hat{x} + \epsilon$ contains the "true" value of the parameter x being estimated with a prespecified probability $1-\alpha$. Therefore, when we make an estimate, we must decide in advance that we would like to be, say, 90% or 95% confident in the estimate. The confidence of interval helps us to know the degree of confidence we have in the estimate. The upper and lower limits of the confidence interval (known as *confidence limits*) are given by

$$\text{upper limit} = \mu + \epsilon \tag{4.42}$$

$$\text{lower limit} = \mu - \epsilon \tag{4.43}$$

where

$$\epsilon = \frac{S\, t_{\alpha/2;N-1}}{\sqrt{N}} \tag{4.44}$$

Thus, if a simulation is performed N times by using different seed values, then in $(1-\alpha)$ cases, the estimate $\hat{\mu}$ lies within the confidence interval and in α cases the estimate lies outside the interval, as illustrated in Figure 4.10. Equation (4.44) provides the error estimate for a given number N of simulation experiments or observations.

If, on the other hand, an accuracy criterion ϵ is prescribed and we want to estimate μ by $\hat{\mu}$ within tolerance of ϵ with at least probability $1-\alpha$, we must ensure that the sample size N satisfies

$$P[|\hat{\mu} - \mu| < \epsilon] \geq 1 - \alpha \tag{4.45}$$

To satisfy this requirement, N must be selected as the small integer satisfying

$$N \geq \left(\frac{S\, t_{\alpha/2;N-1}}{\sqrt{\epsilon}}\right)^2 \tag{4.46}$$

For further discussion on error estimates in simulation, one should consult Merel and Mullin [19] and Chorin [20].

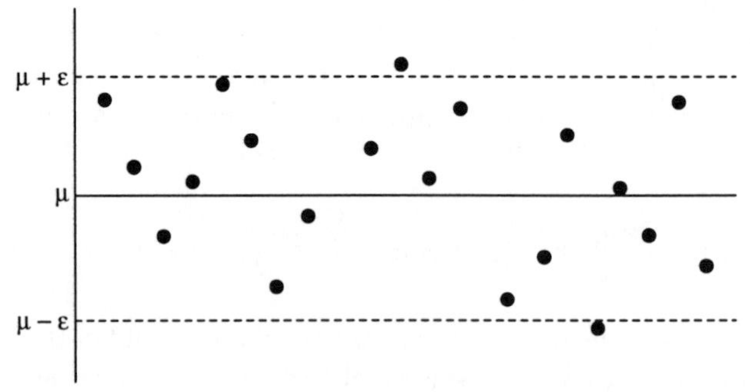

Figure 4.10 Confidence of interval.

Example 4.6

In a simulation experiment, an analyst obtained the mean values of a certain parameter as 7.60, 6.60, 6.97, 7.50, and 7.43 for five simulation runs using different seed values. Calculate the error estimate using a 95% confidence interval.

Solution

We first get the sample mean

$$\mu = \frac{7.60 + 6.60 + 6.97 + 7.50 + 7.43}{5} = 7.22$$

From Eq. (4.31), the sample variance is obtained as

$$S^2 = \frac{(7.60 - 7.22)^2 + \cdots + (7.43 - 7.22)^2}{4} = 0.23793$$

or $S = 0.48778$. Using a 95% confidence interval, $1 - \alpha = 95\%$ (i.e., $\alpha = 0.05$). For five runs ($N = 5$), the t-distribution table gives $t_{\alpha/2;N-1} = 2.776$. Using Eq. (4.44), the error is estimated as

$$\epsilon = \frac{0.48778 \times 2.776}{\sqrt{5}} = 0.6056$$

Thus, the 95% confidence interval for the parameter is

$$\mu - \epsilon < \hat{\mu} < \mu + \epsilon = 6.6144 < \hat{\mu} < 7.8265$$

4.7 Summary

This chapter provides a theoretical basis for the following chapters. It briefly reviews some basic and useful concepts in probability and statistics relating to simulation modeling and analysis. Additional information on these concepts can be obtained from references on probability and statistics [16, 21].

The chapter also discusses procedures for generating pseudorandom numbers and random variates, which are required to simulate stochastic systems. The most common procedures are the multiplicative congruential method and the mixed congruential method. Two common procedures for translating the random numbers into random variates are the inverse transformation method and the rejection method.

Since simulation output is subject to random error, the simulator would like to know how close the point estimate is to the mean value μ it is supposed to estimate. The statistical accurary of the point estimates is measured in terms of the confidence interval. The simulator generates some number of observations, say N, and employs standard statistical methods to obtain the error estimate.

References

[1] H. D. Larsen, *Rinehart Mathematical Tables, Formulas, and Curves.* New York: Rinehart, 1948, pp. 154–155.

[2] R. S. Burington, *Handbook of Mathematical Tables and Formulas.* New York: McGraw-Hill, 1972, pp. 424–427.

[3] W. M. Meredith, *Basic Mathematical and Statistical Tables for Psychology and Education.* New York: McGraw-Hill, 1969, pp. 186–189.

[4] G. S. Fishman, *Concepts and methods in discrete event digital simulation.* New York: John Wiley & Sons, 1973, pp. 167–241.

[5] T. E. Hull and A. R. Dobell, "Random number generators," *SIAM Review*, vol. 4, no. 3, July 1982, pp. 230–254.

[6] D. E. Knuth, *The Art of Computer Programming.* Reading, MA: Addison- Wesley, vol. 2, 1969, pp. 9, 10, 78, 155.

[7] J. Banks and J. Carson, *Discrete-Event System Simulation.* Englewood Cliffs, NJ: Prentice-Hall, 1984, pp. 256–329.

[8] P. A. W. Lewis et al., "A Pseudo-random Number Generator for the System/360", *IBM System Jour.*, vol. 8, no. 2, 1969, pp. 136–146.

[9] S. S. Kuo, *Computer Applications of Numerical Methods.* Reading, MA: Addison-Wesley, 1972, pp. 327–345.

[10] A. M. Law and W. D. Kelton, *Simulation Modeling and Analysis.* New York: McGraw-Hill, 2nd ed., 1991, pp. 462–521.

[11] F. Neelamkavil, *Computer Simulation and Modelling.* Chichester, UK: John Wiley & Sons, 1987, pp. 105–134.

[12] J. R. Emshoff and R. L. Sisson, *Design and Use of Computer Simulation Models.* New York: Macmillan, 1970, pp. 170–182.

[13] H. Kobayashi, *Modeling and Analysis: An Introduction to System Performance Evaluation Methodology.* Reading, MA: Addison-Wesley, 1978, pp. 221–247.

[14] I. M. Sobol, *The Monte Carlo Method.* Chicago: Univ. of Chicago Press, 1974, pp. 24–30.

[15] Y. A. Shreider, *Method of Statistical Testing (Monte Carlo Method).* Amsterdam: Elsevier, 1964, pp. 39–83. Another translation of the same Russian text: Y. A. Shreider, *The Monte Carlo Method (The Method of Statistical Trials).* Oxford: Pergamon, 1966.

[16] H. A. Taha, *Simulation Modeling and SIMNET.* Englewood Cliffs, NJ: Prentice-Hall, 1988, pp. 45–52.

[17] T. H. Naylor et al., *Computer Simulation Techniques.* New York: John Wiley & Sons, 1966, p. 108.

[18] I. S. Sokolinkoff and R. M. Redheffer, *Mathematics of Physics and Modern Engineering.* New York: McGraw-Hill, 1958, pp. 644–649.

[19] M. H. Merel and F. J. Mullin, "Analytic Monte Carlo error analysis," *J. Spacecraft*, vol. 5, no. 11, Nov. 1968, pp. 1304–1308.
[20] A. J. Chorin, "Hermite expansions in Monte-Carlo computation," *J. Comp. Phys.*, vol. 8, 1971, pp. 472–482.
[21] R. B. Cooper, *Introduction to Queueing Theory.* New York: North Holland, 1981, 2nd ed., pp. 34–72.

Problems

4.1 Verify that $f(x) = \frac{1}{2}e^{-|x|}$, $-\infty < x < \infty$, represents a pdf.

4.2 Show that

$$F(x) = \begin{cases} 0, & x < -1 \\ \frac{1}{2}(x+1), & -1 < \quad x < 1 \\ 1, & \text{otherwise} \end{cases}$$

is cdf. Find $f(x)$, $P(-\frac{1}{3} < X < \frac{1}{3})$, $E(x)$, and $V(x)$.

4.3 Show that for the Poisson distribution, the mean and variance are equal.

4.4 The death rate of a hospital is 1/10 per day for a 5-day work week. For a given 5-day week, we begin observing death occurences on Monday. If X is the number of days until the first death occurs, find the probability that: (a) the first week is death free, (b) the first death occurs on Friday of the first week, (c) the first death occurs on Wednesday of the second week.

4.5 Find $P(2 \leq X \leq 5)$ if
(a) X is a normally distributed random variable with parameters $\mu = 2$ and $\sigma = 2$.
(b) X is Poisson distributed with parameter $\gamma = 1$.
(c) X is uniformly distributed over the interval 0.3 to 0.8.

4.6 If X is standard normal random variable, show that (a) $E(X) = 0$ and (b) $V(X) = 1$.

4.7 For a discrete random variable X with probabilities $p_n = P(X = n)$, $n = 0, 1, 2, \ldots$, the moment-generating function is defined as

$$G_x(z) = E(z^x) = \sum_{n=0}^{\infty} p_n z^n$$

(a) Prove that

$$G_x(1) = 1$$

$$\left.\frac{dG_x(z)}{dz}\right|_{z=1} = E(X)$$

$$\left.\frac{d^2G_x(z)}{dz^2}\right|_{z=1} = E(X^2) - E(X)$$

(b) Find $G_x(z)$, $E(X)$, and $V(X)$ for Poisson distribution.

4.8 Repeat Problem 4.7(b) for geometric distribution, where

$$P(X = n) = p_n = pq^{n-1}, \ q = 1 - p$$

4.9 Verify that the mean value and the variance of the gamma distribution are given in Eq. (4.16).

4.10 What are pseudorandom numbers?

4.11 (a) Generate 500 random numbers, exponentially distributed with mean 4, using uniformly distributed random numbers U. Estimate the mean and variance of the variate.
(b) Repeat part (a) using $1 - U$.
(c) Compare your results in parts (a) and (b).

4.12 A uniformly distributed random integers between 11 and 30, inclusive, are to be generated from the random numbers U in the table below. How many of the integers are odd numbers?

0.04493	0.88159
0.52494	0.96119
0.75248	0.63896
0.33824	0.54692
0.45862	0.82391
0.51025	0.23287
0.61962	0.29529
0.79335	0.35963
0.65337	0.15307
0.12472	0.26898

4.13 Suppose U_1, U_2, and U_3 are independent, normally distributed variates. Show that $X_1 = \max(U_1, U_2)$ and $X_2 = \sqrt{U_3}$ have identical distribution. What is the distribution of variate $X_2 = -\ln (U_1 U_2 U_3)$?

4.14 Write a subroutine that will generate a binomial variate. Test the subroutine by generating 100 variates and finding the mean value and variance. Check if your computation agrees with the exact values of $E(X) = mp$ and $V(X) = mpq$. Take $m = 10,\ p = 0.4$.

4.15 Develop a random variate generator for a random variable whose pdf is in Example 4.1.

4.16 Develop a subroutine for computing the mean and variance of a set of random variates. Use the subroutine to confirm the previous problem.

4.17 A random sample of 50 variables taken from a normal population has a mean of 20 and standard deviation 8. Calculate the error with 95% confidence limits.

4.18 In a simulation model of a queueing system, an analyst obtained the mean waiting time for four simulation runs as 42.80, 41.60, 42.48, and 41.80 μs. Calculate the 98% confidence interval for the waiting time.

Chapter 5

Simulation of Token-Passing LANs

The achiever is the only individual who is truly alive.

—George Allen

5.1 Introduction

In this chapter, we discuss simulation of token-passing local area networks (LANs). Both ring and bus topologies are discussed. Token-passing is a technique in which the transmission medium is shared without conflict among LAN users.

In Sections 5.2 and 5.3, we briefly describe the operation of token-passing ring and bus LANs, respectively. Section 5.4 deals with the structure of the simulation model presented in this chapter. Assumptions made in the simulation process and the simulation code are also discussed in this section. A few typical simulation sessions are also included. Section 5.5 summarizes the chapter.

5.2 Operation of Token-Passing Ring LANs

In a token-passing ring local area network, the shared transmission medium is closed on itself and takes the form of a loop as shown in Figure 5.1(a). The information flow is only in one direction. Access to the transmission medium is regulated with the help of a token—a small packet that consists of about eight bits. The token has two possible states: free and busy. A free token indicates that transmission medium is available for transmission; a busy token indicates that the transmission medium is busy.

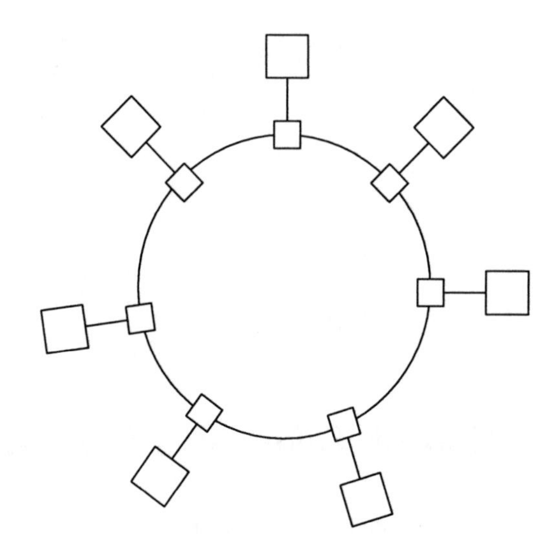

(a) Local area ring network

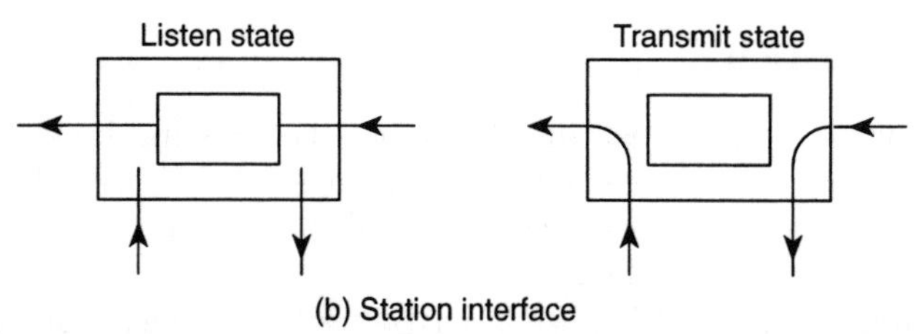

(b) Station interface

Figure 5.1 A typical token-passing local area ring network.

All stations are attached to the transmission medium using active interfaces. By an active interface we mean that the interface can modify the information passing through it. The interface can assume two possible states: listen and transmit. In the listen state, the interface simply monitors the information on the transmission medium. In doing so, it is looking either for a free token to gain access to the transmission medium for transmitting its information, or to identify the information that is addressed to it. In the transmit state, the interface transmits information on one side of the ring and receives information from the other side. These states are shown in Figure 5.1(b). When no station is transmitting information, a free token keeps circulating among all stations in a sequence determined by the physical location of the stations. When a station has some information to transmit, its interface monitors the transmission medium and looks for a free token. As soon as it sees a free token, it captures the token, makes it a busy token (by changing its last bit), and immediately transmits the information. While a station is transmitting, the ring is practically broken at the location of the transmitting station. The station injects information onto the ring from one side and removes information from the other. As the information goes around the ring, other stations monitor the transmission medium identifying the information that belongs to them. As a station finds information addressed to it, the information is copied (not removed) by the station. The transmitting station is responsible for removing its information from the ring. Once a station completes its transmission, it regenerates a free token so that other stations may have opportunity to transmit their information. This process continues, and every station waits for its turn to use the transmission medium.

Once a station captures a free token to gain access to the transmission medium, it can theoretically keep the token for an indefinite period of time. However, this is unacceptable because all stations need a free token to transmit their information. The token must be released (regenerated) by a station after it has used the transmission medium for a certain period of time (called the *token-holding time*).

There are three well-defined schemes, referred to as service disciplines, that can be used to limit the token-holding time in token-passing LANs: exhaustive, gated, and limited-k. In the exhaustive service discipline, once a station captures a free token, it is allowed to use the transmission medium until its buffer becomes empty and the station regenerates the free token so that the next station may capture it to use the transmission medium. In the gated service discipline, a station is allowed to transmit only as many packets of information as were present in its buffer when a free token was captured and then the station regenerates the token. All packets that arrive during the transmission of packets present in the buffer must wait until the next time a free token is captured. In limited-k service discipline, a station is allowed to transmit a maximum of k ($k = 1, 2, 3, \ldots$) packets when it captures a free token. A free token is regenerated either immediately after k packets have been transmitted or when the buffer becomes empty.

Token-holding time also depends upon when a station decides to regenerate a free token. There are three types of operation defined in this regard: multiple-token operation, single-token operation, and single-packet operation. In multiple-token operation, a station regenerates a free token immediately after it has completed the transmission of the last bit of its information. In single-token operation, a station regenerates a free token after it has transmitted the last bit of its information and after it has removed the busy token from the ring. Multiple-token and single- token operations behave in exactly the same way if the packet transmission time is more than the ring latency (the time it takes for a bit to travel around the ring). However, if the packet transmission time is less than the ring latency, the performance of these two operations is quite different. Single-packet operation is a more rigid operation in terms of its requirements for regenerating a free token. In this operation a free token is regenerated only after the last bit of transmitted information has been removed from the ring. The ring must be absolutely free before a free token can be regenerated. Performance comparison of these operations is treated extensively in Hammond and O'Reilly [1].

5.3 Operation of Token-Passing Bus LANs

In a token-passing bus local area network, the shared transmission medium is open ended, as shown in Figure 5.2. Access to the transmission medium is controlled by a token—a small packet that consists of about 48 bits. A station that has the token is allowed to use the transmission medium while other stations wait for their turn.

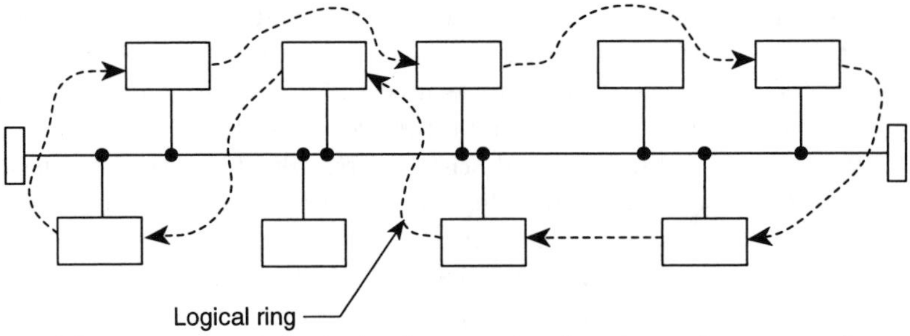

Figure 5.2 A typical token-passing local area bus network.

All stations are attached to the transmission medium by using the passive interfaces. When a station has some information to transmit, it waits for a token. The token is circulated only among those stations that have some information to transmit. Such stations form a logical ring (as opposed to a physical ring). The sequence in which a token is circulated among stations does not depend upon the physical location of the stations as shown in Figure 5.2. When a station completes its transmission, it passes the token to the next station on the logical ring. Once a token is transmitted, it travels away from the station interface in both directions. In order to avoid the possibility of two or more stations capturing the token at the same time, an explicit address of the next station to receive the token is attached to the token. For this reason, the process of token-passing in bus LANs is referred to as explicit token-passing as opposed to the implicit token-passing used in ring LANs.

As only those stations can be part of the logical ring that have some information to transmit, there is an elaborate mechanism for adding stations to the logical ring and removing them from the logical ring. The size of the logical ring varies with the variations in the traffic load of the network. A detailed description of token-passing LANs is given in Hammond and O'Reilly [1] and Keiser [2].

Exhaustive, gated, and limited-k service disciplines as discussed in Section 5.2 are equally applicable to token-passing bus local area networks.

5.4 Simulation Model

In this section, we describe the simulation model developed for simulating token-passing (ring and bus) local area networks. We have used the event-scheduling approach in this simulation and it is self-driven [3, 4]. The program is written in turbo C. Generation of random numbers according to a desired probability distribution is a necessity for any self-driven simulation. In this simulation program we have used a built-in function `rand()` for this purpose. This function generates uniformly distributed numbers, which are then converted to follow a desired probability distribution. In this simulation program, we primarily need exponentially distributed random numbers. The output of the `rand()` function can easily be converted to exponentially distributed numbers using the transform method [4].

A flowchart of the simulation program is given in Figure 5.3 that shows the structure of the program and the logical steps involved. The simulation program has four major sections: initialization, processing, control, and output. In the initialization section, values of the input parameters are read and all the variables are initialized to their appropriate values. This includes initialization of the event list, which contains the timing at which various events are supposed to take place. This section prepares the simulation for execution of events.

The next logical step is to scan the event list and pick an event with the shortest time of occurrence. This process also identifies the event as an arrival of a packet, departure of a packet, or arrival of a token at a station. The program then executes the selected event and updates the values of all the variables affected by the event. This is part of the processing section. After an event has been successfully completed, the program control section checks to see if the simulation should be terminated. If the decision is to continue the simulation, then the event list is scanned again to pick the next event. This process continues until a sufficient number of packets have been transmitted and delivered to their respective destinations to yield reasonably converged simulation results. If the decision is to terminate the simulation, the output section computes the final simulation results and prints them out, and the simulation process stops.

In the next section, we describe the assumptions used in the simulation. Section 5.4.2 describes the input and output variables used in this program. In Section 5.4.3, the simulation program is described in detail, and in Section 5.4.4, some typical simulation results are presented. A brief summary of the chapter is presented in Section 5.5.

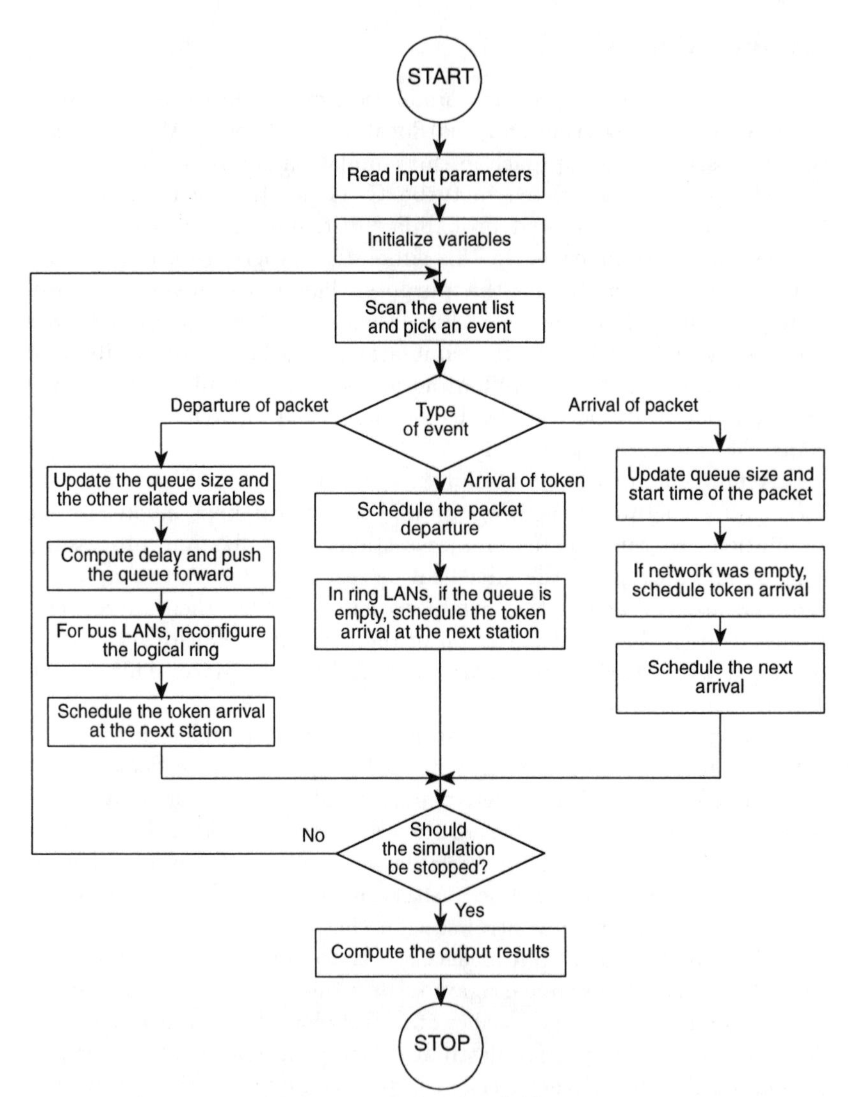

Figure 5.3 Flow chart of the simulation program.

5.4.1 Assumptions

The following assumptions have been made in the simulation process:

- Arrivals at all stations follow a Poisson process.
- All stations generate the traffic at the same rate.
- Packet lengths are fixed.
- Multiple token operation has been used.
- The transmission medium is assumed to be error-free.
- The spacing between stations is the same.
- The source and destination stations are on the average half the ring apart.
- The propagation delay is about five microseconds per kilometer of transmission medium.

5.4.2 Input and Output Variables

Input Variables:

`MAX_STATIONS` Number of stations in the local area network.

`RING_OR_BUS` This variable is used to choose ring or bus local area network for simulation. If `RING_OR_BUS = 1`, the program simulates a ring LAN; otherwise it simulates a bus LAN.

`RATE` Transmission rate of the transmission medium in bits per second. Its reasonable value is 1.0 Mbps to 5.0 Mbps.

`PACKET_LENGTH` Packet length in bits. Its reasonable value is from 500 bits to 10,000 bits.

`MEDIUM_LENGTH` Length of the transmission medium in meters. This is also used to compute the end-to-end propagation delay by assuming that the propagation delay is 5 microseconds per kilometer. A reasonable value of this variable is 1 km to 5 km.

`TOKEN_LENGTH` Token length in bits. Its typical value is 8 bits for ring LANs and 48 bits for bus LANs.

`STN_LATENCY` Station latency of interfaces. For bus LANs, its value is zero; for ring LANs, it is typically 1 bit.

`MAX_PACKETS` The number of packets for which each simulation run is executed before terminating. At lower traffic loads, a value of about 5,000 can be used. However, at higher loads a value of at least 20,000 should be used. Also if the size of the network grows, this value should be increased, too. Basically this parameter is for convergence of simulation results.

FACTOR An accuracy parameter that determines the unit of time for a simulation run. At higher transmission rates this parameter should have a higher value for reasons of accuracy. If **FACTOR** = 1.0, the time unit is in seconds. If **FACTOR** = 1,000, the time unit is milliseconds, and so on. For our simulation the value of **FACTOR** should be at least 1,000.

MAX_Q_SIZE The maximum buffer size at a station.

DEGREES_FR The degrees of freedom to be used for calculating confidence intervals for the output results. In our simulation the range for this variable is from 1 to 10. However, in most of the results, we have used 5 degrees of freedom and a 95% level of confidence.

Output Variables:

average_delay The average delay per packet in seconds per packet.

delay_con_int This variable represents a 95% confidence interval for the average delay with the selected degrees of freedom.

5.4.3 Description of the Simulation Model

In this section we describe the simulation program step-by-step and explain how various aspects of token-passing LANs are simulated. A complete listing of the program is given in Appendix A.

The following segment of the simulation program includes declaration statements indicating constants, real variables, and integer variables. The segment also includes the header files for various built-in functions of the turbo C language used in the simulation process. In this segment, we also assign appropriate values to the input parameters for a simulation run.

```
/* This program simulates token-passing ring and bus LANs.  */


# include     <stdio.h>
# include     <stdlib.h>
# include     <math.h>


# define    MAX_STATIONS     50            /* Maximum number of stations */
# define    RING_OR_BUS      1             /* Flag to choose
                                           ring or bus LAN */
# define    RATE             10000000.0    /* Transmission rate in bps */
# define    PACKET_LENGTH    1000.0        /* Packet length in bits */
# define    MEDIUM_LENGTH    2000.0        /* Medium length in meters */
# define    MAX_PACKETS      10000         /* Maximum packets to be
                                           transmitted in a simulation
                                           run */
# define    FACTOR           1000.0        /* A factor used for changing
                                           units of time */
# define    TOKEN_LENGTH     10.0          /* Token length in bits */
# define    STN_LATENCY      1.0           /* Station latency in bits */
# define    MAX_Q_SIZE       100           /* Maximum queue size */
# define    DEGREES_FR       5             /* Degrees of freedom */
```

```
float arrival_rate; /* Arrival rate (in packets/sec) per station */
float packet_time; /* Packet transmission time */
float stn_latency; /* Station latency in time units */
float token_time; /* Token transmission time */
float tau; /* End-to-end propagation delay */
float t_dist_par[10] = {12.706, 4.303, 3.182, 2.776, 2.571, 2.447, 2.365,
2.306, 2.262, 2.228}; /* T-distribution parameters */
float start_time[MAX_STATIONS][MAX_Q_SIZE]; /* Starting time of packet */
float event_time[MAX_STATIONS][3]; /* Time of occurrence of an event */
float delay_ci[DEGREES_FR + 1]; /* Array to store delay values */

float rho, clock, no_pkts_departed, next_event_time;
float x, logx, flag, infinite, rand_size;
float delay, total_delay, average_delay, walk_time;
float delay_sum, delay_sqr, delay_var, delay_sdv, delay_con_int;


int queue_size[MAX_STATIONS]; /* Current queue size at a station */
int next_stn[MAX_STATIONS]; /* Array to identify the next station */
int previous_stn[MAX_STATIONS]; /* Array to identify previous station */
int in[MAX_STATIONS]; /* Array to identify status of a station */

int i, j, ic, ii, temp_flag, next, next_station, next_event;
int ring_size, ring_or_bus, stn_to_add, temp_stn;
```

The array variables in these statements are explained next.

`start_time[i][j]` The starting time of the packet that is sitting in the buffer of the ith station and occupies jth position; used in calculation of the packet delay when it departs from the station.

`event_time[i][j]` The time at which an event of type j is to occur at the ith station; $j = 0$ implies a packet arrival event, $j = 1$ implies packet departure event, and $j = 2$ implies an arrival event for a free token.

`queue_size[i]` Queue size or buffer occupancy at the ith station.

`next_stn[i]` The (successor) station that is to receive the token after station i has released it.

`previous_stn[i]` The (predecessor) station that is supposed to send a free token to station i.

`t_dist_par[i]` This array contains t-distribution parameters used in calculating confidence intervals. The ith element of this array indicates the parameters to be used in calculating 95% confidence intervals with i degrees of freedom.

`delay_ci[i]` This array is used to temporarily hold the values of delay from various simulation runs with the same input parameters. These values are eventually used in calculating confidence intervals.

`in[i]` This array is used only in token-passing bus LANs. It indicates whether a station is part of logical ring or not. If `in[i]` is 1, that implies station i is a part of the logical ring; and if it is 0, that implies station i is not a part of the logical ring.

The next segment marks the beginning of the main body of the program. The segment starts with some output statements. The variable `arrival_rate` is initialized to zero and its value is incremented within a do-loop. Some of the input variables that are assigned one type of units in the beginning are converted to a more convenient type of units, in this segment. This is done merely for programming convenience. For example, the values of `PACKET_LENGTH`, `TOKEN_LENGTH` and `STN_LATENCY` are initially assigned in bits and are then converted to their equivalent time units (i.e., the time it takes to transmit that many bits). The corresponding variables in time units are `packet_time`, `token_time`, and `stn_latency`, respectively. The variable `tau` represents the end-to-end propagation delay, and its value is calculated assuming that the wave propagation speed on the transmission medium is 5 microseconds per kilometer. The variable `rand_size` represents the largest integer value that the program can handle. This depends on the computer being used. Thus, in order to have the flexibility of running the simulation program on any computer, the program uses `sizeof(int)` function of the turbo C language to determine the size of the integer in bytes. That information is used to determine the value of the variable `rand_size`, which is then used in random number generation.

```
main ()
{
printf("The following results are for:  \n");
printf("Degrees of freedom = % d\n", DEGREES_FR);
printf("Confidence interval = 95 percent \n");
printf(" ========================================== \n");
printf("\n");
arrival_rate        = 0.0;
packet_time         = PACKET_LENGTH * FACTOR / RATE;
stn_latency         = STN_LATENCY * FACTOR / RATE;
token_time          = TOKEN_LENGTH * FACTOR / RATE;
tau                 = MEDIUM_LENGTH * FACTOR * 5.0 * pow
                      (10.0, -9.0);
infinite            = 1.0 * pow (10.0,30.0);
ring_or_bus         = RING_OR_BUS;
rand_size           = 0.5 * pow (2.0, 8.0 * sizeof(int));
```

The next two do-loops are used to run simulation for various values of arrival rates and to conduct several simulation runs (with the same values of input variables but with different random number streams) for the purpose of calculating confidence intervals.

```
for (ii = 0; ii < 10; ii++)
        {
        arrival_rate = arrival_rate + 20.0;
        for (ic = 0; ic <= DEGREES_FR; ic++)
        {
```

In the next segment of the program, we initialize all variables to their appropriate values. The clock for the simulation process is represented by `clock` and is initialized to 0.0. The variable `no_pkts_departed` represents the total number of packets delivered to their destinations and is initialized to 0.0. The variables `total_delay` and `average_delay` are also initialized to 0.0. The variable `flag` represents the state of the network, indicating if some stations have some information packets to transmit (`flag=0`) or if no station has any packets to transmit (`flag=1`). Initially the value of `flag` is 1.0.

```
rho                = 0.0;
clock              = 0.0;
no_pkts_departed   = 0.0;
total_delay        = 0.0;
next_event_time    = 0.0;
average_delay      = 0.0;
flag               = 1.0;
```

The next step is to see if the traffic load is too much for the network to carry. We calculate the traffic intensity `rho` and check to see if it exceeds the network capacity. If it does, the network is assumed to be overloaded, so we terminate the simulation prematurely and notify the user. If it does not exceed the network capacity, the program proceeds to the next step, initializing variables.

```
rho = arrival_rate * PACKET_LENGTH * MAX_STATIONS / RATE;

if (rho >= 1.0)
        {
        printf("Traffic intensity is too high");
        exit(1);
        }
```

In the next do-loop, we initialize the variable `queue_size[i]` to zero for all stations and we also initialize the variable `start_time[i][j]` to zero for all possible entries.

```
for (i = 0; i < MAX_STATIONS; i++)
        {
        queue_size[i] = 0;
        for (j = 0; j < MAX_Q_SIZE; j++)
         {
         start_time[i][j] = 0.0;
         }
        }
```

In the following segment of the simulation program, we initialize some additional variables that are either for token-passing bus LANs or for token-passing ring LAN. Therefore, the program first checks which one of the LANs is being simulated before initializing these variables. Walk time (`walk_time`) is computed separately for ring and bus LANs. It is the time it takes to pass the token from one station to the next station. In ring LANs, it consists of token transmission time, station latency, and the propagation delay from one station to the next station. In bus LANs, station interfaces do not have any latency so the walk time for bus LANs is simply the token transmission time and the propagation delay. The average propagation delay from any station to any other station in bus LANs is assumed to be `tau`/3.0.

We also assign a predecessor **previous_stn[i]** and a successor **next_ stn[i]** for all stations. For ring LANs these remain fixed because the token is always passed onto the next station in the direction of transmission. For bus LANs, the successor and predecessor stations keep changing from time to time. However, for the purpose of initialization, we assume that no station has any information packets to transmit and thus the logical ring size is zero. Therefore, there are no successors or predecessors for any station.

```
if (ring_or_bus == 1)        /* This is for ring LANs.  */
        {
        ring_size = MAX_STATIONS;
        walk_time = token_time + stn_latency + tau/MAX_STATIONS;
        }
else                         /* This is for bus LANs.  */
        {
        ring_size = 0;
        walk_time = token_time + tau/3.0;
        }

for (i = 0; i < MAX_STATIONS; i++)
        {
        if (ring_or_bus == 1)   /* For ring LANs */
         {
         next_stn [MAX_STATIONS-1] = 0;
         previous_stn [0] = MAX_STATIONS-1;
         previous_stn [MAX_STATIONS-1] = MAX_STATIONS-2;
         next_stn [0] = 1;
         if ((i < (MAX_STATIONS-1) && (i > 0)))
          {
          next_stn [i] = i+1;
          previous_stn [i] = i-1;
          }
         }
        else                 /* For bus LANs */
         {
         in[i] = 0;
         next_stn [i] = -1;
         previous_stn [i] = -1;
         }
        }
```

In the following do-loop, the simulation program initializes the event list. All events except the packet arrival events are disabled. This is done because no departure can take place from an empty buffer and buffers will stay empty until some packet arrivals take place. In order to disable an event, a very large value (of the order of 1.0E+30) can be assigned to the event. When the event-list is scanned, the event with the smallest value is selected first. This forces the packet arrivals to take place first, and then we schedule their departures.

```
for (i = 0; i < MAX_STATIONS; i++)
        {
        for (j = 0; j < 3; j++)
         {
         event_time[i][j] = 0.0;
         if (j != 0) event_time[i][j] = infinite;
         }
        }
```

After all the variables have been initialized, the next step is to scan the event list and pick the next event to be processed. The scanning process is merely a process of picking an event with a smallest time value associated with it. The process of scanning the event list, picking the next event, and executing it continues until a sufficient number of packets have been handled by the network. This is the job of the control section of the program. The first statement of the following segment represents the control section. This segment checks to see if a desired number of packets have departed so that the simulation may be stopped. Specifically, it checks to see if the total number of packets delivered (**no_pkts_departed**) is less than the desired maximum (**MAX_PACKETS**). If it is, the program continues the scanning of the event list and picks the next event to process. If not, the program will go to the output section of the program.

```
while (no_pkts_departed < MAX_PACKETS)
         {
         next_event_time = infinite;
         for (i = 0; i < MAX_STATIONS; i++)
          {
          for (j = 0; j < 3; j++)
{
if (next_event_time > event_time[i][j])
        {
           next_event_time = event_time[i][j];
        next_station = i;
           next_event = j;
           }
          }
         }
```

After the event list has been scanned, the following three parameters are produced: **next_station**, **next_event** and **next_event_time**. The variable **next_event_time** represents the time of occurrence of the selected event, which is why it is assigned a very large value before the scanning process begins. The variable **next_station** indicates the station at which the selected event is to take place, and **next_event** denotes the type of the selected event. The value of **next_event** determines which section of the program should execute the selected event.

As the variable **next_event_time** indicates the time of the selected event, the value of **clock** is immediately equated to that of **next_event_time** as shown here.

```
clock = next_event_time;
```

After scanning the event list, if the program does not find a legitimate event (packet arrival, packet departure, or token arrival) to process, it will execute the following short segment. This segment notifies the user that there is some problem with the event list and stops the program. These are some debugging aids that have been built into the program for helping the new users.

```
if (next_event > 2)
        {
        printf("Check the event-list");
        exit(1);
        }
```

If the selected event is a legitimate event, then an appropriate segment is chosen with the help of the following switch statement.

```
switch (next_event)
```

If the selected event happens to be an arrival of a packet (`next_event=0`), the following segment of the program updates the values of all the affected variables of the program and schedules the next packet arrival before scanning the event-list for the next event.

The first thing to do is to see if the selected event is an arrival of packet. If it is (`next_event=0`), this segment is executed; otherwise the program checks to see if some other event has been selected. When a packet arrival occurs, it invariably increases the queue size (`queue_size`) at the selected station (`next_station`) by one. If the arrival causes the queue size to exceed its limit, the program informs the user and terminates. The starting time (`start_time`) of the newly arrived packet is also initialized to the current simulation clock value. This is used in calculating the delay when the packet departs.

If the local area network was empty before the arrival event, all departures and token arrivals must have been disabled (this is done for convenience, for reasons that will be explained shortly). We need to enable token arrivals so that the departure of the packets may be scheduled. Thus, as soon as a packet arrives in an empty LAN, we give the token to the station where packet arrival has taken place. This seems to be an unrealistic assumption, but it does not significantly affect the simulation results. If this assumption is not used, the simulation process becomes too slow at light traffic loads because the probability of a LAN being empty is high at a light traffic load. Also, in an empty LAN, a circulating token creates a large number of events to be executed. These events are less time-consuming (in terms of real time), so the simulation clock does not advance rapidly, and consequently the simulation process slows down.

For token-passing bus LANs, in addition to giving a token to the station where arrival has taken place, we need to reconfigure the logical ring. In the following segment of the program, we have taken all these possibilities into consideration.

```
{
case 0:     /* This is an arrival event.  */
          {
          queue_size[next_station] ++ ;
          if (queue_size[next_station] > MAX_Q_SIZE)
            {
```

```
printf("The queue size is large and is = %d\n",
queue_size[next_station]);
exit(1);
}
if (ring_or_bus == 1) /* This is for a token-passing ring.  */
{
if (flag == 1.0)
 {
 flag = 0.0;
 event_time[next_station][2] = clock;
 }
}
else   /* This is for a token-passing bus.  */
{
 if (flag == 1.0)
  {
  flag = 0.0;
  ring_size = 1;
 in[next_station] = 1;
  next_stn [next_station] = next_station;
  previous_stn [next_station] = next_station;
  event_time[next_station][2] = clock;
  }
 }
```

The first part of the segment is for ring LANs (`ring_or_bus=1`). If the ring is empty, the token is given to the station where arrival of a packet has taken place by scheduling the token arrival event equal to the current value of the simulation clock. The variable **flag** is reset to zero because the network is no longer empty. The second part of the segment is for bus LANs. When an arrival occurs in an empty network, the ring size (`ring_size`) is initialized to 1. There are no successors or predecessors other than the station itself because there is only one station in the logical ring. The token is given to the station where packet arrival has taken place using the same method as discussed for ring LANs.

In the next segment, we schedule the arrival of the next packet at the same station. Packet arrivals are assumed to follow a Poisson process, which means that the packet interarrival times are exponentially distributed. In order to determine when the next arrival will take place, we need exponentially distributed random numbers. We use the `rand()` function for generating uniformly distributed numbers between 0 and 1 and then convert them to exponentially distributed random numbers using the transform method. After scheduling the next arrival event, we go to that part of the program where we check to see if the simulation should be terminated.

```
/* Schedule the next arrival.  */
for (;;)
        {
        x = (float) rand();
        if (x != 0.0) break;
        }
```

```
logx = -log(x/rand_size) * FACTOR / arrival_rate;
event_time[next_station][next_event] = clock + logx;
start_time [next_station][queue_size[next_station]-1] = clock;
break;
}
```

If the event selected after scanning the event list happens to be a departure event (`next_event=1`), the following segment will update all the affected variables. In this segment, we calculate the delay for the departing packet, pass the token to the next station, and if necessary reconfigure the logical ring (in bus LANs), including adjusting successors and predecessors. If the network becomes empty, the packet departure event and token event are also disabled.

First of all we check and make sure that the selected event is a departure event before updating the variables. If the selected event is a departure event (`next_event=1`), processing will start by decrementing the queue size (`queue_size`) at the station by 1. Then the total number of packets departed (`no_pkts_departed`) is incremented by 1, the packet delay is computed by subtracting the packet start time (`start_time`) from the current value of the simulation clock, and the total delay of all packets departed so far is updated (for averaging).

```
case 1:    /* This is a departure event.  */
         {
         queue_size[next_station] -- ;
         no_pkts_departed ++ ;
         delay = clock - start_time [next_station][0];
         total_delay += delay;
```

Once a packet has departed from a station queue, the remaining packets must be pushed forward. This is done only if the queue size is more than zero. As the packets move forward in a queue, they carry their start times with them. The start time of the departing packet is reset to zero. The process of pushing the queue forward at the selected station is done in the following segment of the program. As we are simulating the limited-1 service discipline, we disable departures at the current station and pass the token to the next station.

```
/* Push the queue forward.  */

for (i = 0; i < queue_size[next_station]; i++)
         start_time[next_station][i] = start_time[next_station][i+1];
         start_time[next_station][queue_size[next_station]] = 0.0;
         event_time[next_station][next_event] = infinite;
```

The process of passing the token to the next station differs in ring LANs and bus LANs. If we are dealing with a bus LAN, after a packet departure takes place, we need to give opportunity to the stations that want to join the logical ring and also opportunity to stations that want to leave the logical ring. In this simulation, we have assumed that a station that has some information packets to transmit and is not currently part of the logical ring may become part of the ring only if its address is between the current station and the current successor to the current station. If more than one station satisfies this criterion, then the first (closest) one is selected to join the logical ring. The following segment of the simulation program selects a station to be added to the logical ring. After a station has been selected, the logical ring size is incremented by 1 and the successor and predecessors are, of course, adjusted accordingly. If no station satisfies the criterion, the logical ring remains unchanged.

```
if (ring_or_bus == 0) /* For bus LANs */
        {
        stn_to_add = -1;
        for (i = next_station + 1; i < MAX_STATIONS; i++)
         {
        if ((queue_size[i] > 0) && (in[i] == 0)) stn_to_add = i;
         if (stn_to_add != -1) continue;
         }
        if (stn_to_add == -1)
         {
         for (i = 0; i < next_station - 1; i++)
          {
         if ((queue_size [i] > 0) && (in[i] == 0)) stn_to_add = i;
          if (stn_to_add != -1) continue;
          }
         }
if (stn_to_add != -1)
{
temp_stn = next_stn[next_station];
next_stn[next_station] = stn_to_add;
next_stn[stn_to_add] = temp_stn;
previous_stn[stn_to_add] = next_station;
previous_stn[temp_stn] = stn_to_add;
ring_size ++ ;
in[stn_to_add] = 1;
}
```

If the station queue has become empty, after the departure of a packet, then the station must leave the logical ring. If this condition is satisfied at a station, it is deleted from the logical ring, the logical ring size is decremented by 1, and the successors and predecessors are adjusted accordingly. If the logical ring size becomes zero as a result of this deletion of a station, the variable `flag` is set to 1 and token passing is disabled until the next packet arrival takes place. In all other cases, the token arrival is scheduled at the next station (successor).

```
if (queue_size[next_station] == 0)
        {
        ring_size -- ;
        in[next_station] = 0;
         if (ring_size == 0)
           {
         next_stn[next_station] = -1;
         previous_stn[next_station] = -1;
           flag = 1.0;
           }
         else
           {
          next = next_stn[next_station];
          event_time[next][2] = clock + walk_time;
          next_stn[previous_stn[next_station]] =
           next_stn[next_station];
          previous_stn[next] = previous_stn[next_station];
```

```
            }
        }
        else
        {
        next = next_stn[next_station];
        event_time[next][2] = clock + walk_time;
        }
      }
```

In case of a ring LAN, the ring size does not vary with departures or arrivals. However, due to the limited-1 service discipline, the token is passed to the next station after every departure. Once the token has passed through all the stations, the program checks if all stations are empty. If they are, the variable **flag** is set to 1 and the process of token passing is disabled until a packet arrival takes place. In all other cases, the token arrival is scheduled at the next station. Finally, we go to that part of the program where we check if the simulation should be terminated.

```
if (ring_or_bus == 1)    /* For ring LANs */
      {
      next = next_stn[next_station];
      if ((next == 0) && (queue_size[next_station] == 0))
       {
       temp_flag = 1;
       for (i = 0; i < MAX_STATIONS; i++)
        {
        if (queue_size[i] != 0)
         {
         event_time[next][2] = clock + walk_time;
         temp_flag = 0;
         break;
         }
        }
       if (temp_flag == 1)
         {
         flag = 1.0;
         event_time[next][2] = infinite;
         }
        }
      else
       {
       event_time[next][2] = clock + walk_time;
       }
      }
break;
}
```

The next segment of the program deals with the process of token passing among stations. This process is slightly different for bus and ring LANs. In ring LANs, the token goes from one station to the next station in the direction of transmission. In bus LANs, the token is passed to the next station (successor) according to a sequence determined by the logical ring. When a token is passed to the next station in a ring LAN, that station may or may not have any information packets to transmit. If it does not have any information to transmit, the token is passed to the next station and the process is repeated. However, if the station does have some information packets to transmit, it captures the token, makes it busy and starts its transmission. Upon completing transmission, the station regenerates a free token and passes it to the next station. On the other hand, in bus LANs, the token goes only to those stations that have some information packets to transmit. Once a station receives a token, it transmits its information, regenerates a new token, and passes it to the next station. If none of the stations have any information to transmit, the token passing event is disabled until the next arrival takes place at some station.

The first item before this segment is processed is to check if this event is really a token arrival event (`next_event=2`). If it is, we need to schedule the departure of a packet from this station if it has any. Thus, if the queue size at this station is larger than zero, departure of a packet is scheduled. Checking the queue size is necessary only for ring LANs because the token goes to all stations irrespective of their queue size. In bus LANs, however, if the token arrives at a station and the station does not have any packet to transmit, that implies that there is some discrepancy in the token passing process. In such a case, the program informs the user about the anomaly and stops the simulation. In the case of ring LANs, we schedule the token arrival at the next station and quit this segment. If we notice that the token has completed a circulation around the ring, the program checks if all the stations are empty. If so, the token-passing process is disabled and the programs wait until the next packet arrival at some station.

```
case 2:      /* This is a token arrival event.  */
          {
          event_time[next_station][2] = infinite;
          if (queue_size[next_station] > 0)
           {
           event_time[next_station][1] = clock + packet_time;
           }
          else
           {
          if (ring_or_bus == 0) /* For bus LANs */
           {
           printf("There is something wrong (bus LAN)");
           }
          else /* For ring LANs */
            {
            next = next_stn[next_station];
           if ((next == 0) && (queue_size[next] == 0))
             {
             temp_flag = 1;
             for (i = 0; i < MAX_STATIONS; i++)
              {
              if (queue_size[i] != 0)
               {
               event_time[next][2] = clock + walk_time;
               temp_flag = 0;
               break;
               }
              }
```

```
                if (temp_flag == 1)
                  {
                  flag = 1.0;
                  event_time[next][2] = infinite;
                  }
                }
               else
                {
                event_time[next][2] = clock + walk_time;
                }
              }
            }
          break;
          }
        }
      }
```

Once it has been decided that simulation is to be terminated, the program computes the average delay per packet for the current simulation run. This is computed by dividing the total delay of all packets (`total_delay`) by the total number of packets delivered (`no_pkts_departed`). It is also divided by the variable `FACTOR` to convert the average delay to seconds. This segment of the program stores the value of delay in an array (`delay_ci`) until a specified number of simulation runs have been made for calculating the confidence interval for the average delay per packet.

```
average_delay = total_delay / (no_pkts_departed * FACTOR);
delay_ci[ic] = average_delay
}
```

Once all the simulation runs have been completed, the following segment of the simulation program computes the average delay (`delay_sum`) and 95% confidence interval (`delay_con_int`) for the average delay. The output results are then printed before the simulation program stops.

```
delay_sum = 0.0;
delay_sqr = 0.0;
for (ic = 0; ic <= DEGREES_FR; ic++)
      {
      delay_sum += delay_ci[ic];
      delay_sqr += pow (delay_ci[ic],2.0);
      }
delay_sum = delay_sum / (DEGREES_FR + 1);
delay_sqr = delay_sqr / (DEGREES_FR + 1);
delay_var = delay_sqr - pow(delay_sum,2.0);
delay_sdv = sqrt(delay_var);
delay_con_int = delay_sdv * t_dist_par[DEGREES_FR-1]/sqrt (DEGREES_FR);
printf("For an arrival rate = %g\n",arrival_rate);
printf("The average delay = %g", delay_sum);
printf(" +- %g\n", delay_con_int);
printf("\n");
      }
  }
```

One of the do-loops increments the value of `arrival_rate` (by 20.0) and repeats the simulation process for a specified number of times. Several more output results can be easily computed from this program.

Some of these are delay histograms, average queue size, number of token circulations per second, and average size of the logical ring in bus LANs.

5.4.4 Typical Simulation Sessions

In this section, we present typical simulation runs for token passing ring as well as bus LANs. The appropriate input values are declared as constants at the beginning of the simulation program. The arrival rate per station is automatically incremented within the program. The program is then compiled and executed by giving an appropriate command for the computer being used.

In order to illustrate the simulation of a token-passing bus LAN, the following set of values may be assigned to the input parameters:

```
MAX_STATIONS = 50
RING_OR_BUS = 0 /* This is a token-passing bus LAN. */
RATE = 10000000.0
PACKET_LENGTH = 1000.0
MEDIUM_LENGTH = 2000.0
MAX_PACKETS = 10000
FACTOR = 1000.0
TOKEN_LENGTH = 50.0
STN_LATENCY = 0.0 /* Not used in bus LANs */
MAX_Q_SIZE = 100
DEGREES_FR = 5
```

The output results of the simulation program for the given set of input parameters are as follows:

```
The following results are for:
Degrees of freedom = 5
Confidence interval = 95 percent
==========================================

For an arrival rate = 20
The average delay = 0.000122978 +- 1.22711e-06

For an arrival rate = 40
The average delay = 0.000133929 +- 1.45604e-06

For an arrival rate = 60
The average delay = 0.000151054 +- 2.1269e-06

For an arrival rate = 80
The average delay = 0.000167302 +- 3.52123e-06

For an arrival rate = 100
The average delay = 0.000194165 +- 5.48323e-06

For an arrival rate = 120
The average delay = 0.000242528 +- 3.19022e-06

For an arrival rate = 140
The average delay = 0.000339794 +- 1.73057e-05

For an arrival rate = 160
The average delay = 0.000532814 +- 3.09825e-05

For an arrival rate = 180
The average delay = 0.00201972 +- 0.00066182
```

For simulating a token-passing ring LAN, the following set of values may be assigned to the input parameters:

```
MAX_STATIONS = 50
RING_OR_BUS = 1 /* This is a token-passing ring LAN. */
RATE = 10000000.0
PACKET_LENGTH = 1000.0
MEDIUM_LENGTH = 2000.0
MAX_PACKETS = 10000
FACTOR = 1000.0
TOKEN_LENGTH = 10.0
STN_LATENCY = 1.0 /* Used only in ring LANs */
MAX_Q_SIZE = 100
DEGREES_FR = 5
```

The output results of the simulation program for the given set of input parameters are as follows:

```
The following results are for:
Degrees of freedom = 5
Confidence interval = 95 percent
==========================================

For an arrival rate = 20
The average delay = 0.000124185 +- 1.26429e-06

For an arrival rate = 40
The average delay = 0.000138816 +- 9.81389e-07

For an arrival rate = 60
The average delay = 0.000159987 +- 2.30818e-06

For an arrival rate = 80
The average delay = 0.000182707 +- 2.92595e-06

For an arrival rate = 100
The average delay = 0.000216262 +- 4.70304e-06

For an arrival rate = 120
The average delay = 0.000271338 +- 4.20807e-06

For an arrival rate = 140
The average delay = 0.000368605 +- 1.50873e-05

For an arrival rate = 160
The average delay = 0.000520035 +- 1.42912e-05

For an arrival rate = 180
The average delay = 0.00103385 +- 0.000137447
```

These results are summarized in Table 5.1.

5.5 Summary

In this chapter we have discussed simulation of token-passing ring and bus local area networks. After briefly describing the operation of token passing in ring and bus LANs, we have explained a simulation program step by step that takes the input parameters from the user and simulates one of the LANs at a time to a desired level of convergence. The results are presented in terms of average delay per packet with 95% confidence interval.

References

[1] J. L. Hammond and J.P. O'Reilly, *Performance Analysis of Local Computer Networks*, Reading, MA; Addison-Wesley 1986.

[2] G. E. Keiser, *Local Area Networks*, New York; McGraw-Hill, 1989.
[3] M. Ilyas and H. T. Mouftah, *Simulation Tools for Computer Communication Networks*, Conference Record of the IEEE GLOBECOM '88, 1988, pp. 1702–1706.
[4] H. Kobayashi, *Modeling and Analysis: An Introduction to System Performance Evaluation Methodology*, Reading, MA; Addison-Wesley, 1978.

Problems

1. Modify the simulation program to find out the average number of token circulations per unit time in token-passing ring as well as in token-passing bus LANs.
2. Modify the simulation program to find out the distribution of the logical ring size in token-passing bus LANs.
3. The amount of time a LAN spends in activities other than actual transmission of user information is called overhead. Modify the simulation programs to compute the overhead time in both token passing bus and ring LANs.
4. One of the nice things about developing a simulation model by yourself is that you can obtain any desired performance parameters. One such performance parameter is distribution of delay. Modify the simulation program to obtain the delay distribution.

Table 5.1 Simulation results for a token-passing LAN.

Arrival Rate (in packets per second)	Average Delay per Packet (token-passing bus LAN)	Average Delay per Packet (token-passing ring LAN)
20.0	0.000123 ± 1.2271E−06	0.000124 ± 1.2643E−06
40.0	0.000134 ± 1.4560E−06	0.000139 ± 9.8139E−07
60.0	0.000151 ± 2.1269E−06	0.000160 ± 2.3081E−06
80.0	0.000167 ± 3.5212E−06	0.000182 ± 2.9260E−06
100.0	0.000194 ± 5.4832E−06	0.000216 ± 4.7030E−06
120.0	0.000243 ± 3.1902E−06	0.000271 ± 4.2080E−06
140.0	0.000340 ± 1.7305E−05	0.000368 ± 1.5087E−05
160.0	0.000533 ± 3.0982E−05	0.000520 ± 1.4291E−05
180.0	0.002020 ± 6.6182E−04	0.001034 ± 1.3745E−04

Chapter 6

Simulation of CSMA/CD LANs

An optimist sees an opportunity in every calamity. A pessimist sees a calamity in every opportunity.

—Herbert V. Prochnow

6.1 Introduction

One popular and widely used local area network (LAN) is Ethernet. It is based upon a multiple access protocol called CSMA/CD: Carrier Sense Multiple Access with Collision Detection. In this technique, all stations attached to the LAN make their transmission decisions on the basis of the status of the transmission medium (busy or idle) as they see it at their interfaces. The transmissions initiated by various stations may overlap and may cause a collision. This necessitates retransmission of all collided packets of information. In this chapter, we discuss simulation of LANs that use CSMA/CD as an access protocol.

In Section 6.2 we briefly describe the operation of local area networks that use the CSMA/CD access protocol. A few of its variations are also described. Section 6.3 describes the structure of the simulation model presented in this chapter. Assumptions made in the simulation process and the simulation code are also discussed in this section. Some typical simulation results are also included. Section 6.4 summarizes the contents of this chapter.

6.2 Operation of CSMA/CD LANs

In CSMA/CD-based local area bus networks, the transmission medium is open ended. All stations are attached to the transmission medium using passive interfaces as shown in Figure 6.1. Access to the transmission medium is decided solely by the stations that are attempting to transmit.

Before transmitting its information packets, a station senses the state of the transmission medium to see if it is already busy (in transporting information packets) or idle. If a station finds the medium busy at its interface, then the transmission of its information packet must be delayed. However, if the medium is sensed free at its interface, the transmission of information packets may proceed.

Although each station transmits only when it senses the transmission as free, a collision with other transmissions may still take place. This is because a transmission decision is made only on the basis of local information (i.e., the information that an interface sees at its interface), not on the basis of the overall situation of the transmission medium. When a station transmits its packets, it takes a small but finite amount of time (end-to-end propagation delay τ seconds, in the worst case) before this information reaches all stations. Consider a situation in which a station senses the transmission medium at its interface as free and begins its transmission. Another station senses the transmission medium at its interface as free because the previous station's transmission has not reached its interface yet, so this station also begins its transmission. Now two transmissions are in progress at the same time and will definitely collide within τ seconds.

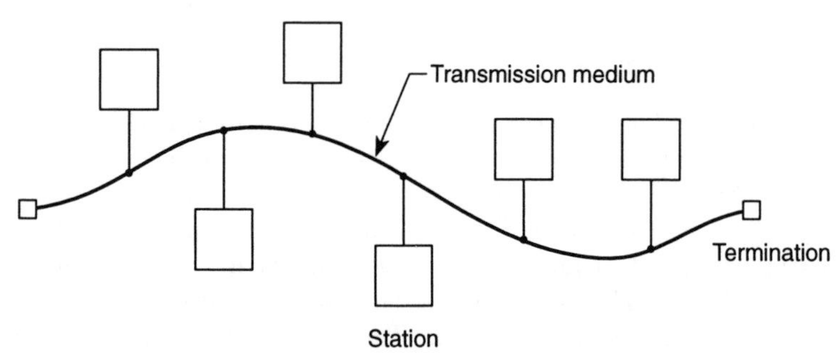

Figure 6.1 A typical local area bus network.

During transmission of their information packets, all stations monitor transmitted information packets on the transmission medium. The information on the medium is compared with what each station is transmitting. If these two match then it is assumed that the transmission is successful. However, if the transmitted information differs from what is on the transmission medium, it is assumed that a collision has taken place and a retransmission is needed.

In CSMA/CD-based LANs, when a collision has been detected, all transmitting stations abort their transmission. The station that first detects a collision starts transmitting a collision enforcement signal also known as *jamming signal.* This is to inform all the stations that a collision has taken place and they should wait until after the jamming signal is over and a predefined silence period has elapsed. After the silence period, business starts as usual. A typical sequence of events in a successful and an unsuccessful transmission is shown in Figure 6.2.

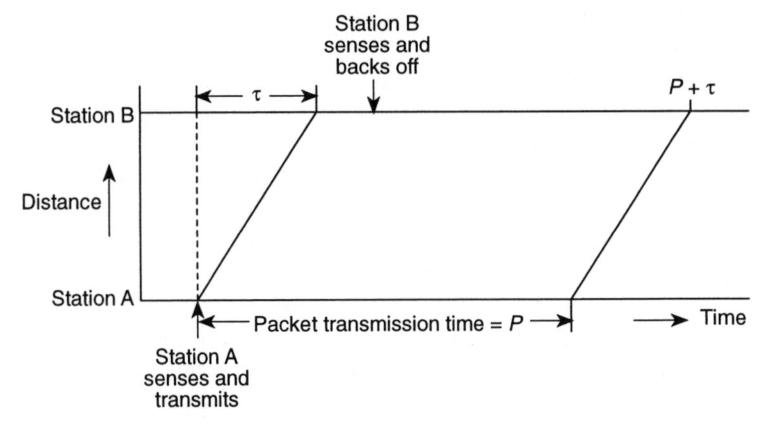

(a) Successful transmission in a CSMA/CD LAN.

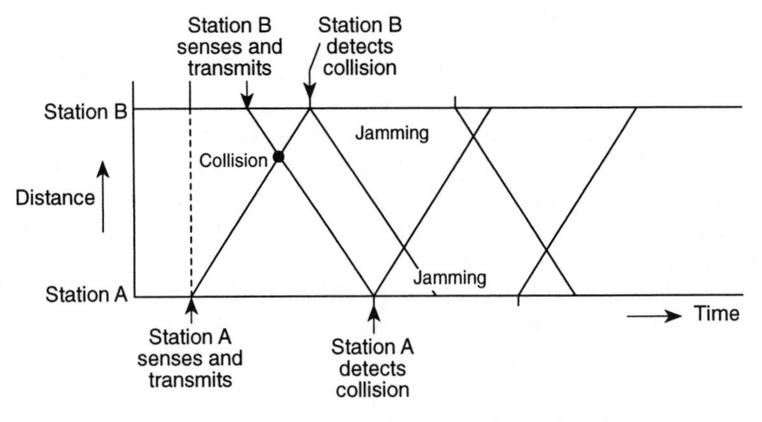

(b) Unsuccessful transmission in a CSMA/CD LAN.

Figure 6.2 Operation of CSMA/CD local area networks.

6.2.1 Nonpersistent and p-Persistent CSMA/CD

As mentioned earlier, the decision about transmission of a packet is made solely by a station and the decision depends upon the status of the transmission medium as seen by the point of interface. The decision may vary slightly depending upon which version of CSMA/CD is being used even if the status of the transmission medium is the same.

Let us assume that a station has an information packet to transmit. It senses the transmission medium at its interface and finds it free. In the case of nonpersistent CSMA/CD, the station will definitely transmit its packet. However, in the case of p-persistent CSMA/CD, the packet is transmitted with probability p, and the transmission is delayed by τ seconds (end-to- end propagation delay) with probability $(1 - p)$. If p happens to equal 1, as is the case in Ethernet implementations, the packet will also be transmitted immediately after the transmission medium is sensed free.

On the other hand, if the transmission medium is sensed busy, no packet should be transmitted. In the case of nonpersistent CSMA/CD, the station backs off and senses the transmission medium again after a random duration of time. In the case of p-persistent CSMA/CD, the station keeps on checking the transmission medium continuously until it becomes free. As soon as the medium becomes free, the station transmits its packet with probability p and delays it by τ seconds with probability $(1 - p)$. Obviously, if $p = 1$ in the p-persistent case, the station will immediately transmit after the transmission medium becomes free. In this case, if more than one station was checking the transmission medium at the same time, all of them will transmit almost at the same time and will collide with probability 1.

6.3 Simulation Model

In this section, we describe the model developed for simulating CSMA/-CD local area networks. The simulation program is written in turbo C and is based upon the event-scheduling approach. It is self-driven [1, 2] and needs generation of random numbers according to a desired probability distribution function. For this purpose, we have used a built-in function `rand()`. This function generates uniformly distributed numbers, which are then converted to follow a desired probability distribution. In this simulation program, we primarily need exponentially distributed random numbers. The output of the `rand()` function can easily be converted to exponentially distributed numbers using the transform method [2].

A flowchart of the simulation program given in Figure 6.3 shows the structure of the program and the logical steps involved. The simulation program has four major sections: initialization, processing, control, and output.

In the initialization section, values of the input parameters are read and all the variables are initialized to their appropriate values. This includes initialization of the event list that contains the timing at which various events are supposed to take place. This section prepares the simulation for execution of events.

The next logical step is to scan the event list and pick an event with the shortest time of occurrence. This process also identifies the event as an arrival of a packet at a station, a transmission attempt at a station, a collision of packets, or a departure of a packet at a station. After selecting an event, the program executes the selected event and updates the values of all the variables affected by the event. After an event has been successfully completed, the control section of the program checks to see if the simulation should be terminated. If the decision is to continue the simulation, then the event list is scanned again to pick the next event.

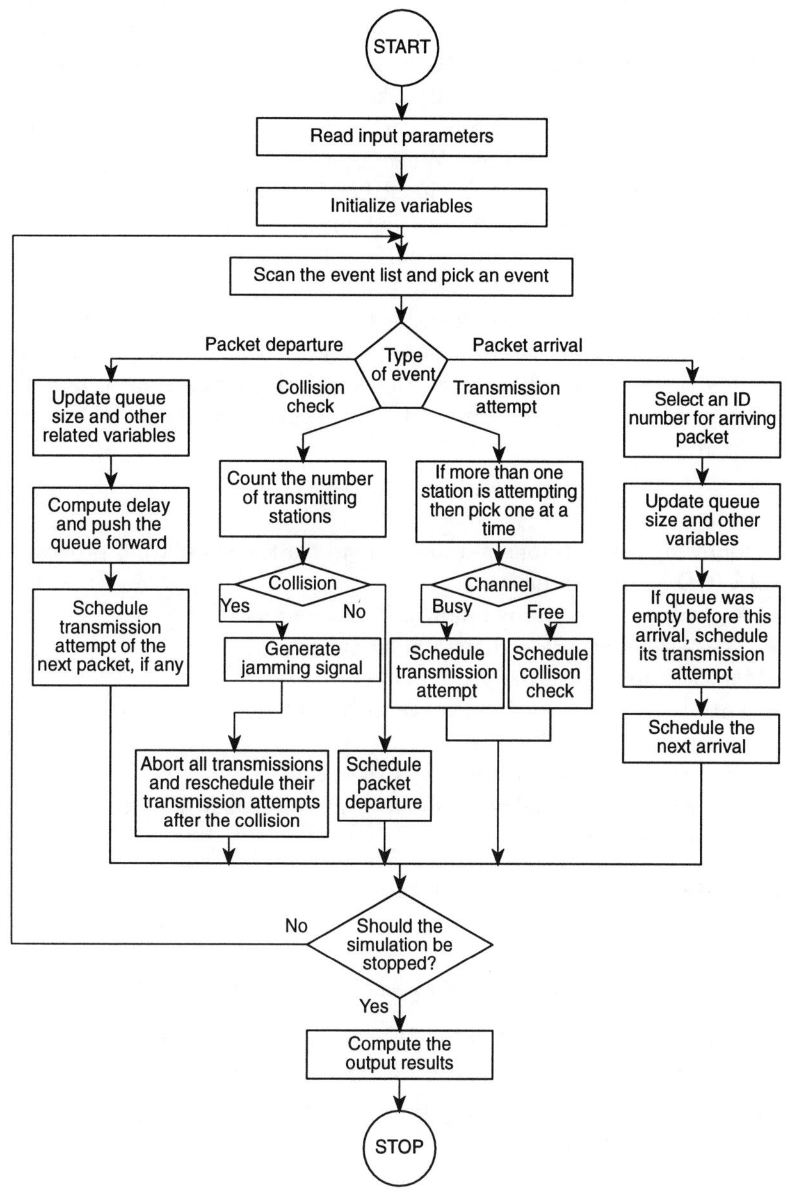

Figure 6.3 Flow chart of the simulation program.

This process continues until enough packets have been transmitted and delivered to their respective destinations to yield reasonably converged simulation results. If the decision is to terminate the simulation, the output section computes the final simulation results and prints them out.

One novel aspect of this simulation program is that each packet carries with it a unique identification number. The number is assigned to a packet as soon as it arrives at station and remains with the packet until the packet departs from the network. This identification number is in all calculations related to a packet. For example, the packet delay is calculated by using the simulation clock value at the departure time and the start time of a packet. When a packet is departing, we can find its start time by looking at its identification number and the starting time associated with that identification number. This approach becomes very helpful when there is a large number of parameters associated with a packet.

In the next section, we describe the assumptions used in the simulation. All the input and output variables used in this program are described in Section 6.3.2. In Section 6.3.3 the simulation program is described in detail. In Section 6.3.4, some typical simulation results are presented.

6.3.1 Assumptions

The following assumptions have been made in the simulation process of CSMA/CD local area networks:

- Arrivals at all stations follow a Poisson process.
- All stations generate the traffic at the same rate.
- Packet lengths are fixed.
- The transmission medium is assumed to be error-free, and any errors are only due to collisions.
- The spacing between stations is the same.
- The propagation delay is about 5 microseconds per kilometer of transmission medium.

6.3.2 Input and Output Variables

Input variables:

`MAX_STATIONS` Number of stations in the local area network.

`BUS_RATE` Transmission rate of the transmission medium in bits per second. A reasonable value is from 1.0 Mbps to 10.0 Mbps.

`PACKET_LENGTH` Packet length in bits. Its reasonable value is from 500 bits to 10,000 bits.

`BUS_LENGTH` Length of the transmission medium in meters. This is also used to compute the end-to-end propagation delay by assuming that the propagation delay is 5 microseconds per kilometer. A reasonable value for this variable is 1 km to 5 km.

`MAX_BACKOFF` Maximum length of the backoff interval in terms of number of slots. A reasonable value for this variable is 5.0 to 20.0.

`PERSIST` The persistent parameter. This may vary, from 0 to 1 (inclusive). If it is zero, it represents nonpersistent CSMA/CD; other values represent p-persistent CSMA/CD.

`JAM_PERIOD` The duration of the jamming signal in terms of number of slots. A reasonable value for this variable is 5.0 to 20.0.

`MAX_PACKETS` The number of packets for which each simulation run is executed before terminating. At lower traffic loads a value of about 5,000 can be used. However, at higher loads a value of at least 20,000 should be used. Also, if the size of the network grows, this value should be increased, too. Basically, this parameter is for convergence of simulation results.

`FACTOR` An accuracy parameter that determines the unit of time for a simulation run. At higher transmission rates this parameter should have a higher value for the reasons of accuracy. If `FACTOR` = 1.0, the time unit is in seconds. If `FACTOR` = 1,000, the time unit is milliseconds, and so on. For our simulation the value of `FACTOR` should be at least 1,000.

`MAX_Q_SIZE` The maximum buffer size at a station. Its reasonable value is from 100 to 500.

`ID_SIZE` The size of the identification array. Its reasonable value is about 5,000.

`DEGREES_FR` The number of degrees of freedom to be used for calculating confidence intervals for the output results. In our simulation the range for this variable is from 1 to 10. However, in most of the results, we have used 5 degrees of freedom and 95% level of confidence.

Output Variables:

`average_delay` The average delay per packet in seconds per packet.

`delay_con_int` The 95% confidence interval for the average delay with the selected degrees of freedom.

`utilization` The fraction of time the transmission medium is utilized.

`utilization_con_int` The 95% confidence interval for the utilization with a given number of degrees of freedom.

`throughput` The average throughput in terms of packets per second.

`collision_rate` The collision rate in terms of collisions per second.

6.3.3 Description of the Simulation Model

In this Section, we describe the simulation program step by step and explain how various aspects of a CSMA/CD-based local area network are simulated. A complete listing of the simulation program is given in Appendix B.

The following segment of the simulation program includes declaration statements indicating constants, real variables, and integer variables. The segment also includes the header files for various built-in functions of the turbo C language used in the simulation process. We also assign appropriate values to the input parameters for a simulation run.

```
/* This program simulates CSMA/CD local area networks.  */

# include     <stdio.h>
# include     <stdlib.h>
# include     <math.h>


# define    MAX_STATIONS      10           /* Number of stations */
# define    BUS_RATE          2000000.0    /* Transmission rate in bps */
# define    PACKET_LENGTH     1000.0       /* Packet length in bits */
# define    BUS_LENGTH        2000.0       /* Bus length in meters */
# define    MAX_BACKOFF       15.0         /* Backoff period in slots */
# define    PERSIST           0.5          /* Persistence */
# define    JAM_PERIOD        5.0          /* Jamming period */
# define    MAX_PACKETS       10000        /* Maximum packets to be trans-
                                           mitted in a simulation run */
# define    FACTOR            1000.0       /* A factor used for changing
                                           units of time */
# define    MAX_Q_SIZE        500          /* Maximum queue size */
# define    ID_SIZE           5000         /* Size of the identity array */
# define    DEGREES_FR        5            /* Degrees of freedom */


float arrival_rate; /* Arrival rate (in packets/sec) per station */
float arrival_rate_slots; /* Arrival rate (in packets/slot) per station */
float packet_time; /* Packet transmission time */
float t_dist_par[10] = {12.706, 4.303, 3.182, 2.776, 2.571, 2.447, 2.365,
                  2.306, 2.262, 2.228}; /* T-distribution parameters */
float start_time[ID_SIZE]; /* Starting time of packet */
float event_time[MAX_STATIONS][4]; /* Time of occurrence of an event */
float delay_ci[DEGREES_FR+1]; /* Array to store delay values */
float utilization_ci[DEGREES_FR+1]; /* Array for utilization values */
float throughput_ci[DEGREES_FR+1]; /* Array to store throughput values */
float collision_rate_ci[DEGREES_FR+1]; /* Array to save collision rates */


float slot_size, p, ch_busy;
float rho, clock, d_clock, no_pkts_departed, next_event_time;
float x, logx, rand_size, infinite;
float delay, total_delay, average_delay;
float delay_sum, delay_sqr, delay_var, delay_sdv, delay_con_int;
float utilization, utilization_sum, utilization_sqr;
float utilization_var, utilization_sdv, utilization_con_int;
float throughput, throughput_sum;
float collision_rate, collision_rate_sum, collision_end_time;
float select_prob, backoff_time, packet_slots;
```

```
int queue_size[MAX_STATIONS]; /* Current queue size at a station */
int int queue_id[MAX_STATIONS][MAX_Q_SIZE]; /* Array for packet ID's */
int id_list[ID_SIZE]; /* Array of id_numbers */
int id_attempt_stn[MAX_STATIONS]; /* Array for attempting stations */

int i, j, ic, ii, next_station, next_event, next, id_number;
int no_attempts, no_trans, no_collisions, select_flag;
```

The array variables in these statements are explained next:

`start_time[i]` The starting time of the packet whose identification number is used in calculation of the packet delay when the packet departs from the station.

`event_time[i][j]` The time at which an event of type j is to occur at the ith station; $j = 0$ implies a packet arrival event, $j = 1$ implies a transmission attempt, $j = 2$ represents a collision check event, and $j = 3$ indicates a departure event.

`queue_size[i]` The queue size or buffer occupancy at the ith station.

`id_list[i]` This variable is used to assign an identification number to each packet in the network. It is assumed that there cannot be more than `ID_SIZE` packets in the network at any given time.

`queue_id[i][j]` The identification number of the packet that is sitting in the buffer of the ith station and occupies the jth position. It is assumed that the buffer size dose not exceed `MAX_Q_SIZE`.

`t_dist_par[i]` This array contains T-distribution parameters used in calculation of confidence intervals. The ith element of this array indicates the parameters to be used in calculating 95% confidence intervals with i degrees of freedom.

`delay_ci[i]` This array is used to temporarily hold the values of delay from various simulation runs with the same input parameters. These values are eventually used in calculating confidence intervals.

`utilization_ci[i]` This array is used to temporarily hold the values of utilization from various simulation runs with the same input parameters. These values are eventually used in calculating confidence intervals.

`throughput_ci[i]` This array is used to temporarily hold the values of throughput from various simulation runs with the same input parameters. These values may eventually be used in calculation of confidence intervals.

collision_rate_ci[i] This array is used to temporarily hold the values of collision rate from various simulation runs with the same input parameters. These values may eventually be used in calculating confidence intervals.

The next segment marks the beginning of the main body of the simulation program. The variable arrival_rate is initialized to zero, and its value is incremented within a do-loop. Some of the input variables that are assigned one type of unit in the beginning are converted to a convenient type of units in this segment. This is done merely for programming convenience. For example, the value of variable PACKET_LENGTH is initially assigned in bits and is then converted to its equivalent time units. The corresponding variable in time units is packet_time. The variable slot_size represents the end-to-end propagation delay, and its value is calculated assuming that the wave propagation speed on the transmission medium is 5 microseconds per kilometer. The variable rand_size represents the largest integer value that the program can handle. This depends upon the computer being used. Thus, in order to have the flexibility of running the simulation program on any computer, the program uses the sizeof(int) function of the turbo C language to determine of the size of integer in bytes. That information is used to determine the value of the variable rand-size, which is then used in random number generation.

```
main ()
{
printf("The following results are for:  \n");
printf("Degrees of freedom = %d\n", DEGREES_FR);
printf("Confidence Interval = 95 percent \n");
printf(" ======================================== \n");
printf("\n");

arrival_rate = 0.0;
slot_size = BUS_LENGTH * FACTOR * 5.0 * pow (10.0, -9.0);
p = PERSIST;
packet_time = PACKET_LENGTH * FACTOR / BUS_RATE;
packet_slots = (float) (int) (packet_time/slot_size) + 1.0;
infinite = 1.0 * pow (10.0, 30.0);
rand_size = 0.5 * pow (2.0, 8.0 * sizeof(int));
```

The next two do-loops are used to run simulation for various values of arrival rates and to conduct several simulation runs (with the same values of input variables but with different random number streams) for the purpose of calculating confidence intervals.

```
for (ii=0; ii < 10; ii++)
        {
        arrival_rate = arrival_rate + 20.0;
        for (ic = 0; ic <= DEGREES_FR; ic++)
        {
```

In the next segment of the program, we initialize all variables to their appropriate values. The clock for the simulation process is represented by `clock` and is initialized to 0.0. Another clock, `d_clock`, is also initialized to 0.0. This clock represents the timing epochs at which all events other than packet arrival, take place. This clock is needed because we are assuming slotted operation. The variable `utilization` represents utilization of the transmission medium. It is computed by summing all time durations in which the transmission medium is utilized and dividing this sum by the simulation clock value at the end of simulation. This is initialized to 0.0 at the beginning of each simulation process. The variable `ch_busy` represents the status of the channel (0.0 for idle and 1.0 for busy). As the channel is initially idle, this variable is initialized to 0.0. The variable `no_pkts_departed` represents the total number of packets delivered to their destinations and is initialized to 0.0. The variable `no_collisions` represents the total number of collisions and is initialized to 0.0. Similarly, the variable `collision_end_time` marks the ending time of a collision. As there is no collision at the start, this variable is also initialized to 0.0. The variables `total_delay` and `average_delay` are also initialized to zero.

```
rho                   = 0.0;
ch_busy               = 0.0;
clock                 = 0.0;
d_clock               = 0.0;
collision_end_time    = 0.0;
utilization           = 0.0;
no_pkts_departed      = 0.0;
total_delay           = 0.0;
next_event_time       = 0.0;
average_delay         = 0.0;
no_collisions         = 0;
select_flag           = 0;
```

The next step is to see if the traffic load is too much for the network to carry. We calculate the traffic intensity `rho` and check to see if it exceeds the network capacity. If it does, the network is assumed to be overloaded so we terminate the simulation prematurely and notify the user. If it does not exceed the network capacity, the program proceeds to the next step, initializing variables.

```
rho = arrival_rate * PACKET_LENGTH * MAX_STATIONS / BUS_RATE;

if (rho >= 1.0)
{
printf("Traffic intensity is too high");
exit(1);
}
```

In the following do-loops, we initialize the variable `queue_size[i]` to zero for all stations. We also initialize the variables `start_time[i][j]`, `queue_id[i][j]`, and `id_list[i]` to zero for all possible entries. The variable `arrival_rate_slots` represents the arrival rate per slot and is also computed in the following segment.

```
arrival_rate_slots = arrival_rate * slot_size;
     for (i = 0; i < MAX_STATIONS; i++) queue_size[i]=0;
     for (i = 0; i < ID_SIZE; i++)
       {
```

```
        start_time[i] = 0.0;
        id_list[i] = 0;
        }
    for (i = 0; i < MAX_STATIONS; i++)
        {
        for(j = 0; j < MAX_Q_SIZE; j++) queue_id[i][j]=0;
    }
```

In the next do-loop, the simulation program initializes the event list. All events except the packet arrival events are disabled because no departure can take place from an empty buffer, and buffers will stay empty until some packet arrivals take place. In order to disable an event, a very large value (of the order of 1.0E+30) can be assigned to the event. When the event list is scanned, the event with the smallest value is selected first. This forces the packet arrivals to take place first, and then we schedule their departures.

```
for (i = 0; i < MAX_STATIONS; i++)
        {
        for (j = 0; j < 4; j++)
         {
         event_time[i][j] = infinite;
         x = (float) rand();
         x = x * FACTOR/rand_size;
         if (j == 0) event_time[i][j] = x;
         }
        }
```

After all the variables have been initialized, the next step is to scan the event list and pick the next event to be processed. The scanning process is merely a process of picking the event associated with the smallest time value. The process of scanning the event list, picking the next event, and executing it continues until a sufficient number of packets have been handled by the network. This is the job of the control section of the program. The first statement of the following segment represents the control section. This segment checks if a desired number of packets have departed so that the simulation may be stopped. Specifically, it checks to see if the total number of packets delivered (`no_pkts_departed`) is less than the desired maximum (`MAX_PACKETS`). If it is the program continues the scanning of the event list and picks the next event to process. If not, the program will go to the output section of the program.

```
while (no_pkts_departed < MAX_PACKETS)
        {
        next_event_time = infinite;
        for (i = 0; i < MAX_STATIONS; i++)
         {
         for (j = 0; j < 4; j++)
          {
          if (next_event_time > event_time[i][j])
           {
           next_event_time = event_time[i][j];
           next_station = i;
           next_event = j;
```

```
            }
        }
    }
```

After the event list has been scanned, the following three parameters are produced: **next_station**, **next_event** and **next_event _time**. The variable **next_event_time** represents the time of occurrence of the selected event, which is why it is assigned a very large value before the scanning process begins. The variable **next_station** indicates the station at which the selected event is to take place, and **next_event** denotes the type of the selected event. The value of next_event determines which section of the program should execute the selected event.

As the variable **next_event_time** indicates the time of the selected event, the value of **clock** is immediately equated to that of **next_event _time** as shown here:

```
clock = next_event_time;
```

After scanning the event list, if the program does not find a legitimate event (packet arrival, packet departure, or token arrival) to process, it will execute the following short segment. This segment notifies the user that there is some problem with the event list and stops the program. These are some debugging aids that have been built into the program for helping the new users.

```
if (next_event > 3)
        {
        printf("Check the event_list");
        exit(1);
        }
```

If the selected event is a legitimate event, then an appropriate segment is chosen with the help of the following switch statement. However, before doing that, we update the value of **d_clock** as shown in the following segment of the simulation program.

```
while (d_clock <= clock) d_clock ++ ;
switch (next_event)
```

If the selected event happens to be an arrival of a packet (**next_event =0**), the following segment of the program updates the values of all the affected variables of the program and schedules the next packet arrival before scanning the event list again for the next event.

The first thing to do is to see if the selected event is an arrival of packet. If it is (**next_event=0**), this segment is executed, otherwise the program checks to see if some other event has been selected. When a packet arrival occurs, we select an identification number from the ID list. If all identification numbers have already been used, the program informs the user and stops the simulation process. If an identification number is available, the queue size (**queue_size**) at the selected station (**next_station**) is incremented by one. The start time (**start_time**) of the newly arrived packet is also initialized to the current simulation clock value, and the packet is placed at the appropriate place in the queue. If this arrival causes the queue size to exceed its limit, the program informs the user and stops the simulation.

If the station's queue was empty before this arrival, its transmission attempt event is scheduled according to the value of **d_clock**. However, if a collision has recently occurred and a period of silence is still to come, the transmission attempt event of the newly arrived packet is scheduled to be at the end of the collision period.

```
{
case 0:  /* This is an arrival event.  */
{

    /* Select an identification for the arriving message */
    id_number = -1;
    for (i = 0; i < ID_SIZE; i++)
      {
      if (id_list[i] == 0)
        {
        id_number = i;
        id_list[i] = 1;
        break;
        }
      if (id_number != -1) continue;
      }
    if (id_number == -1)
      {
      printf("Check the ID list.");
      exit(1);
      }
    queue_size[next_station] ++ ;
    if (queue_size[next_station] > MAX_Q_SIZE)
      {
      printf("The queue size is large and is = %d\n",
      queue_size[next_station]);
      exit(1);
      }
    queue_id[next_station][(queue_size[next_station]-1)] =
    id_number;
    start_time[id_number] = clock;
    if (queue_size[next_station] == 1)
      {
      event_time[next_station][1] = d_clock;
      if (event_time[next_station][1] <= collision_end_time)
      event_time[next_station][1] = collision_end_time + 1.0;
      }
```

In the next small segment, we schedule the arrival of the next packet at the same station. Packet arrivals are assumed to follow a Poisson process, which means that the packet interarrival times are exponentially distributed. In order to determine when the next arrival will take place, we need exponentially distributed random numbers. We use the function **rand()x** to generate uniformly distributed numbers and then convert them to exponentially distributed random numbers using the transform method. After scheduling the next arrival event, we go to that part of the program, where we check to see if the simulation should be terminated.

```
/* Schedule the next arrival */
for (;;)
        {
        x = (float) rand();
        if (x != 0.0) break;
        }
logx = -log(x/rand_size) * FACTOR / arrival_rate_slots;
event_time[next_station][next_event] = clock + logx;
break;
}
```

If, after the event list is scanned the selected event happens to be an event representing a transmission event (**next_event=1**), the following segment will update all the affected variables. Otherwise, we check to see if some other event has been selected.

In the event of a transmission attempt, first of all we realize that there may be more than one station ready to attempt transmission at the same time. However, the scanning process will always pick up a station represented by the lowest value of **next_station** and that is obviously an unfair process. In order to avoid this, we check to see how many stations are ready to attempt transmission at the same time as that of the selected event. If this number is more than one, then we pick one of the stations at random to attempt a transmission.

```
case 1:  /* This is an attempt event.  */
    {
    no_attempts = 0;
    for (i = 0; i < MAX_STATIONS; i++)
      {
      if (event_time[i][1] == clock)
        {
        no_attempts ++ ;
        id_attempt_stn[no_attempts - 1] = i;
        }
      }
    select_flag = 0;
    if (no_attempts > 1)
      {
      x = (float) rand();
      x = x/rand_size;
      for (i = 0; i < no_attempts; i++)
        {
        select_prob = (float) (i+1)/ ((float) no_attempts);
        if (x <= select_prob)
          {
          next_station = id_attempt_stn[i];
          select_flag = 1;
          }
        if (select_flag == 1) continue;
        }
      }
```

Once a station has been selected, we schedule either its collision detection event or another transmission attempt depending upon the status of the transmission medium and whether the network is using a nonpersistent or p-persistent CSMA/CD access mechanism. If the transmission medium is free (ch_busy $= 0.0$) and the network is using nonpersistent version of CSMA/CD protocol ($p = 0.0$), the packet transmission begins and we schedule a collision detection event after one slot time. On the other hand, if the medium is free and the network is using a p-persistent CSMA/CD protocol (p is not zero), we decide either to transmit or not to transmit based upon the value of parameter p. For this purpose, we generate a uniformly distributed random number between 0 and 1, and compare it with the value of p. If the random number is less than p then the packet transmission begins and we schedule a collision detection event after one slot. However, if the random variable is greater than p, a new transmission attempt is scheduled after a duration of one slot time. This is implemented in the following segment of the simulation program.

```
if (ch_busy == 0.0)
    {
    if (p == 0.0)
      {
      event_time[next_station][2] = clock + 1.0;
      event_time[next_station][1] = infinite;
      }
    else
      {
      x = (float) rand();
      x = x/rand_size;
      if (x < p)
        {
        event_time[next_station][2] = clock + 1.0;
        event_time[next_station][1] = infinite;
        }
      else
        {
        event_time[next_station][1] = clock + 1.0;
        if (event_time[next_station][1] <= collision_end_time)
        event_time[next_station][1] = collision_end_time + 1.0;
        event_time[next_station][2] = infinite;
        }
      }
    }
```

If the transmission is sensed busy (ch_busy $= 1.0$) and the network is using a nonpersistent CSMA/CD protocol ($p = 0.0$), we reschedule another transmission attempt after a random duration of time. This random duration cannot exceed MAX_BACKOFF slots. On the other hand, if the transmission medium is busy (ch_busy $= 1.0$), we reschedule another transmission attempt either at the end of the current activity of the transmission medium or one slot after the transmission medium becomes free, depending upon the value of the parameter p. This is again implemented by generating a uniformly distributed random number between 0 and 1 and comparing it with the value of parameter p. This is implemented in the following segment of the simulation program. After the scheduling of a collision event or another transmission attempt has been completed, we go to that part of the program, where we check to see if the simulation should be terminated.

```
if (ch_busy == 1.0)
    {
    if (p == 0.0)
      {
      x = (float) rand();
      x = x/rand_size;
      backoff_time = (float) (int) (x * MAX_BACKOFF);
      if (backoff_time < 1.0) backoff_time = 1.0;
      event_time[next_station][1] = clock + backoff_time;
      if (event_time[next_station][1] <= collision_end_time)
      event_time[next_station][1] = collision_end_time +
      backoff_time;
      event_time[next_station][2] = infinite;
      }
    else
      {
      event_time[next_station][1] = clock + 1.0;
      if (event_time[next_station][1] <= collision_end_time)
      event_time[next_station][1] = collision_end_time + 1.0;
      event_time[next_station][2] = infinite;
      }
    }
break;
}
```

In this section, the program checks to see if a collision has taken place. The first item is to make sure that the selected event is that of a collision check (**next_event** $= 2$). If it is, we proceed to update all the affected variables. Initially, we count the number of stations (**no_trans**) transmitting at the same time. If **no_trans** is greater than 1, then there is a collision. In this case, all transmitting stations must back off, send a jamming signal for a specified duration of time (**JAM_PERIOD**), wait for a silence period equal to two slot times, and schedule their transmission attempts after a random duration of time. All stations that were attempting to transmit before the collision occurred must also back off and reschedule their transmission attempts later. However, if there is no collision (**no_trans** $= 1$), the transmission is successful and the simulation program schedules departure of the packet. After scheduling the departure, the simulation program checks to see if the simulation is to be terminated.

```
case 2:  /* This is a transmission event, */
    {
    no_trans = 0;
    for (i = 0; i < MAX_STATIONS; i++)
    if (event_time[i][2] == clock) no_trans ++ ;
    if (no_trans > 1)
      {
        {
        collision_end_time = clock + JAM_PERIOD + 2.0;
        no_collisions ++ ;
        }
      for (i = 0; i < MAX_STATIONS; i++)
        {
        if (event_time[i][2] == clock)
```

```
            {
            event_time[i][2] = infinite;
            x = (float) rand();
            x = x/rand_size;
            backoff_time = (float) (int) (x * MAX_BACKOFF);
            if (backoff_time < 1.0) backoff_time = 1.0;
            event_time[i][1] = collision_end_time + backoff_time;
            }
          if (event_time[i][1] <= collision_end_time)
            {
            x = (float) rand();
            x = x/rand_size;
            backoff_time = (float) (int) (x * MAX_BACKOFF);
            if (backoff_time < 1.0) backoff_time = 1.0;
            event_time[i][1] = collision_end_time + backoff_time;
            }
          }
        }
      else
        {
        if (ch_busy != 1.0)
          {
          event_time[next_station][3] = clock + packet_slots ;
          event_time[next_station][2] = infinite;
          ch_busy = 1.0;
          }
        else
          {
          if (p == 0.0)
            {
            x = (float) rand();
            x = x/rand_size;
            backoff_time = (float) (int) (x * MAX_BACKOFF);
            if (backoff_time < 1.0) backoff_time = 1.0;
            event_time[next_station][1] = clock + backoff_time;
            if (event_time[next_station][1] <= collision_end_time)
            event_time[next_station][1] = collision_end_time +
            backoff_time;
            event_time[next_station][2] = infinite;
            }
          else
            {
            event_time[next_station][1] = clock + 1.0;
            if (event_time[next_station][1] <= collision_end_time)
            event_time[next_station][1] = collision_end_time + 1.0;
            event_time[next_station][2] = infinite;
            }
          }
        }
      break;
      }
```

If the event selected after the event list is scanned happens to be a departure event, then the following section of the simulation program updates all the affected variables and statistics. This is the segment where we calculate the delay of the departing packet.

First of all, we check to make sure that the selected event is a departure event. If it is a departure event (**next_event** = 3), the processing of this segment starts by checking the identification number (**id_number**) of the departing packet. This number is used in recalling other parameters associated with the departing packet. Then we make the transmission medium idle and decrement the queue size (`queue_size`) by one at the selected station. Then the remaining packets in the queue are pushed forward. As the packets move forward in a queue, they carry their identification numbers with them. All this is done by the following segment of the simulation program.

```
case 3:  /* This is a departure event.  */
     {
     id_number = queue_id[next_station][0];
     ch_busy = 0.0;
     queue_size[next_station] -- ;

     /* Push the queue forward.  */

     for (i = 0; i < queue_size[next_station]; i++)
       queue_id[next_station][i] = queue_id[next_station][i+1];
     queue_id[next_station][queue_size[next_station]] = 0;
```

As a packet departs from a station, the program calculates its delay and updates other statistics about the network. The delay is calculated by subtracting the starting time of the packet from the current value of the simulation clock. The starting time (`start_time`) of the departing packet is associated with its identification number (`id_number`). The total delay of all packets that have departed so far is also updated for the purpose of computing the average delay at the end of the simulation. The program also increments the total number of departed packets by one so as to compute the average delay per packet at the end of the simulation. The identification number of the departing packet is released so that it can be used by another packet. The program also updates the total utilization (`utilization`) of the transmission medium which is used in computing the average utilization before the simulation ends.

```
delay = clock - start_time[id_number];
total_delay += delay;
id_list[id_number] = 0;
no_pkts_departed += 1.0;
utilization += packet_slots;
```

Finally, before leaving the processing of the departure event, we schedule the next event for transmission attempt by the station. This event is scheduled only if the departing packet does not leave queue as empty. After that the program checks if the simulation is to be terminated.

```
event_time[next_station][3] = infinite;
if (queue_size[next_station] > 0)
     {
     event_time[next_station][1] = clock + 1.0;
     if (event_time[next_station][1] <= collision_end_time)
     event_time[next_station][1] = collision_end_time + 1.0;
     }
else
```

```
        {
        event_time[next_station][1] = infinite;
        event_time[next_station][2] = infinite;
        }
        break;
        }
      }
    }
```

Once it has been decided that the simulation is to be terminated, the program processes the following segment of the program. In this segment, the program computes the average utilization of the transmission medium, the average delay per packet, the average throughput, and the average collision rate. The average utilization is computed by dividing the duration of time (`utilization`) for which the transmission medium was used for successful packet transmissions by the total simulation time. The total simulation time is represented by the last value of the simulation clock (`clock`). The average delay is computed by dividing the total delay of all packets (`total_delay`) by the number of packets successfully delivered (`no_pkts_departed`). The average throughput is computed by dividing the total number of departed packets (`no_pkts_departed`) by the total simulation time (`clock`). Similarly, the average collision rate is computed by dividing the total number of collisions (`no_collisions`) by the total simulation time (`clock`). Some of these output results are converted to their appropriate units by taking `FACTOR` and slot size (`slot_size`) into consideration. These results are temporarily stored in their respective arrays until a specified number of simulation runs have been made for calculating a confidence interval for some of the results.

```
utilization = utilization / clock;
average_delay = total_delay * slot_size / (no_pkts_departed * FACTOR);
throughput = no_pkts_departed * FACTOR / (clock * slot_size);
collision_rate = (float) no_collisions * FACTOR / (clock * slot_size);
utilization_ci[ic] = utilization;
delay_ci[ic] = average_delay;
throughput_ci[ic] = throughput;
collision_rate_ci[ic] = collision_rate;
}
```

Once all the simulation runs have been completed, the following segment of the simulation program computes the averages and 95% confidence interval for various output results. The output results are then printed before the simulation program stops.

```
delay_sum = 0.0;
delay_sqr = 0.0;
utilization_sum = 0.0;
utilization_sqr = 0.0;
throughput_sum = 0.0;
collision_rate_sum = 0.0;
for (ic = 0; ic <= DEGREES_FR; ic++)
    {
```

```
            delay_sum += delay_ci[ic];
            delay_sqr += pow (delay_ci[ic],2.0);
            utilization_sum += utilization_ci[ic];
            utilization_sqr += pow (utilization_ci[ic],2.0);
            throughput_sum += throughput_ci[ic];
            collision_rate_sum += collision_rate_ci[ic];
            }
        delay_sum = delay_sum / (DEGREES_FR + 1);
        delay_sqr = delay_sqr / (DEGREES_FR + 1);
        delay_var = delay_sqr - pow(delay_sum,2.0);
        delay_sdv = sqrt(delay_var);
        delay_con_int = delay_sdv * t_dist_par[DEGREES_FR-1]/sqrt(DEGREES_FR);
        utilization_sum = utilization_sum / (DEGREES_FR + 1);
        utilization_sqr = utilization_sqr / (DEGREES_FR + 1);
        utilization_var = utilization_sqr -pow(utilization_sum,2.0);
        utilization_sdv = sqrt(utilization_var);
        utilization_con_int = utilization_sdv *
              t_dist_par[DEGREES_FR-1]/sqrt(DEGREES_FR);
        throughput_sum =
        throughput_sum / (DEGREES_FR + 1);
        collision_rate_sum = collision_rate_sum / (DEGREES_FR + 1);
        printf("For an arrival rate = %g\n",arrival_rate);
        printf("The average delay = %g", delay_sum);
        printf(" +- %g\n", delay_con_int);
        printf("The utilization = %g", utilization_sum);
        printf(" +-%g\n", utilization_con_int);
        printf("The throughput = %g\n", throughput_sum);
        printf("The collision rate = %g\n", collision_rate_sum);
        printf("\n");
    }
}
```

There are many more output results that can be easily computed by using this simulation program. These results may include delay histograms, delay distributions, and average queue size. It is left to the reader to explore the potentials of this simulation program.

6.3.4 Typical Simulation Sessions

Three sets of simulation results are presented in this section. The first set is for nonpersistent CSMA/CD (p = 0.0), the next two sets are for *p*-persistent CSMA/CD, one with p = 0.1 and the other with p = 0.5.

For the first set of results, the input parameters are assigned the following values:

```
MAX_STATIONS = 10
BUS_RATE = 2000000.0
PACKET_LENGTH = 1000.0
BUS_LENGTH = 2000.0
MAX_BACKOFF = 15.0
PERSIST = 0.0 /* Nonpersistent */
JAM_PERIOD = 5.0
MAX_PACKETS = 5000
FACTOR = 1000.0
MAX_Q_SIZE = 500
ID_SIZE = 5000
DEGREES_FR = 5
```

The output results of the simulation program for the given set of input parameters are as follows:

```
The following results are for:
Degrees of freedom = 5
Confidence Interval = 95 percent
==========================================

For an arrival rate = 20
The average delay = 0.000562744 +- 4.17349e-06
The utilization = 0.101816 +- 0.0033588
The throughput = 199.639
The collision rate = 0.329983

For an arrival rate = 40
The average delay = 0.000606906 +- 8.21336e-06
The utilization = 0.205386 +- 0.00509104
The throughput = 402.718
The collision rate = 2.27824

For an arrival rate = 60
The average delay = 0.000672506 +- 8.85559e-06
The utilization = 0.308417 +- 0.00508099
The throughput = 604.739
The collision rate = 6.54644

For an arrival rate = 80
The average delay = 0.000763181 +- 2.71142e-05
The utilization = 0.406525 +- 0.00860557
The throughput = 797.109
The collision rate = 18.5004

For an arrival rate = 100
The average delay = 0.000904571 +- 2.27698e-05
The utilization = 0.517648 +- 0.0131853
The throughput = 1015
The collision rate = 41.6712

For an arrival rate = 120
The average delay = 0.00109228 +- 7.00992e-05
The utilization = 0.609048 +- 0.0193935
The throughput = 1194.21
The collision rate = 78.1405

For an arrival rate = 140
The average delay = 0.00170232 +- 0.000126584
The utilization = 0.717276 +- 0.0091623
The throughput = 1406.42
The collision rate = 215.309

For an arrival rate = 160
The average delay = 0.00378275 +- 0.00136432
The utilization = 0.811852 +- 0.0131487
The throughput = 1591.87
The collision rate = 573.405

For an arrival rate = 180
The average delay = 0.0543311 +- 0.0170496
The utilization = 0.832368 +- 0.00826609
The throughput = 1632.09
The collision rate = 2218.93
```

These results are summarized in Table 6.1.

Table 6.1 Simulation results.

Arrival Rate	Average Delay	Utilization	Through-put	Collision Rate
20.0	0.000562 ± 4.17E−06	0.102 ± 0.00336	199.64	0.329
40.0	0.000606 ± 8.21E−06	0.205 ± 0.00509	402.71	2.278
60.0	0.000672 ± 8.85E−06	0.308 ± 0.00508	604.73	6.546
80.0	0.000763 ± 2.71E−05	0.406 ± 0.00860	797.10	18.50
100.0	0.000904 ± 2.27E−05	0.517 ± 0.01318	1015.0	41.67
120.0	0.001092 ± 7.01E−05	0.609 ± 0.01939	1194.2	78.14
140.0	0.001702 ± 0.000126	0.717 ± 0.00916	1406.4	215.3
160.0	0.003782 ± 0.001364	0.812 ± 0.01314	1591.9	573.4
180.0	0.054331 ± 0.017049	0.832 ± 0.00826	1632.1	2218.9

For simulating *p*-persistent CSMA/CD LANs, one may give the following set of input parameters.

```
MAX_STATIONS = 10
BUS_RATE = 2000000.0
PACKET_LENGTH = 1000.0
BUS_LENGTH = 2000.0
MAX_BACKOFF = 15.0
PERSIST = 0.1 /* p-Persistent */
JAM_PERIOD = 5.0
MAX_PACKETS = 5000
FACTOR = 1000.0
MAX_Q_SIZE = 500
ID_SIZE = 5000
DEGREES_FR = 5
```

The output results for this set of parameters are as follows:

```
The following results are for:
Degrees of freedom = 5
Confidence Interval = 95 percent
========================================

For an arrival rate = 20
The average delay = 0.000658616 +- 3.04385e-06
The utilization = 0.101 +- 0.00172139
The throughput = 198.04
The collision rate = 0.379955

For an arrival rate = 40
The average delay = 0.000715458 +- 4.15925e-06
The utilization = 0.20489 +- 0.00567377
The throughput = 401.745
The collision rate = 1.37544
```

```
For an arrival rate = 60
The average delay = 0.000791979 +- 2.07238e-05
The utilization = 0.310382 +- 0.00900632
The throughput = 608.591
The collision rate = 4.93771

For an arrival rate = 80
The average delay = 0.00088773 +- 2.11368e-05
The utilization = 0.407861 +- 0.00824132
The throughput = 799.727
The collision rate = 10.0247

For an arrival rate = 100
The average delay = 0.0010505 +- 1.68148e-05
The utilization = 0.514659 +- 0.0100435
The throughput = 1009.14
The collision rate = 25.4778

For an arrival rate = 120
The average delay = 0.00136187 +- 8.72094e-05
The utilization = 0.612764 +- 0.0141284
The throughput = 1201.5
The collision rate = 56.2565

For an arrival rate = 140
The average delay = 0.00194209 +- 0.00034533
The utilization = 0.714333 +- 0.0237597
The throughput = 1400.65
The collision rate = 113.977

For an arrival rate = 160
The average delay = 0.00366124 +- 0.000784148
The utilization = 0.811327 +- 0.0126383
The throughput = 1590.84
The collision rate = 254.407

For an arrival rate = 180
The average delay = 0.0270767 +- 0.00771216
The utilization = 0.886182 +- 0.00501053
The throughput = 1737.61
The collision rate = 864.478
```

These results are summarized in the following table:

Arrival Rate	Average Delay	Utilization	Through-put	Collision Rate
20.0	0.000658 ± 3.04E−06	0.101 ± 0.00172	198.04	0.380
40.0	0.000715 ± 4.16E−06	0.205 ± 0.00567	401.74	1.375
60.0	0.000791 ± 2.07E−06	0.310 ± 0.00901	608.59	4.937
80.0	0.000887 ± 2.11E−05	0.408 ± 0.00824	799.72	10.02
100.0	0.001050 ± 1.68E−05	0.515 ± 0.01004	1009.1	25.47
120.0	0.001362 ± 8.72E−05	0.612 ± 0.01412	1201.5	56.25
140.0	0.001942 ± 0.000345	0.714 ± 0.02376	1400.65	113.9
160.0	0.003661 ± 0.000784	0.811 ± 0.01263	1590.8	254.5
180.0	0.027077 ± 0.007712	0.886 ± 0.00501	1737.6	864.5

For still another illustration of simulating a *p*-persistent CSMA/CD LAN, we give the following set of input parameters.

```
MAX_STATIONS = 10
BUS_RATE = 2000000.0
PACKET_LENGTH = 1000.0
BUS_LENGTH = 2000.0
MAX_BACKOFF = 15.0
PERSIST = 0.5 /* p-Persistent */
JAM_PERIOD = 5.0
MAX_PACKETS = 5000
FACTOR = 1000.0
MAX_Q_SIZE = 500
ID_SIZE = 5000
DEGREES_FR = 5
```

The output results for this set of parameters are as follows:

```
The following results are for:
Degrees of freedom = 5
Confidence Interval = 95 percent
=========================================

For an arrival rate = 20
The average delay = 0.000570462 +- 4.44283e-06
The utilization = 0.10224 +- 0.00292783
The throughput = 200.47
The collision rate = 0.667547

For an arrival rate = 40
The average delay = 0.000616678 +- 9.42117e-06
The utilization = 0.202414 +- 0.00448696
The throughput = 396.889
The collision rate = 5.06478

For an arrival rate = 60
The average delay = 0.000669938 +- 1.16101e-05
The utilization = 0.305902 +- 0.00749913
The throughput = 599.807
The collision rate = 15.6545

For an arrival rate = 80
The average delay = 0.000782643 +- 1.57005e-05
The utilization = 0.418693 +- 0.0088699
The throughput = 820.968
The collision rate = 52.1089

For an arrival rate = 100
The average delay = 0.000936336 +- 4.76193e-05
The utilization = 0.516871 +- 0.00857569
The throughput = 1013.47
The collision rate = 115.336

For an arrival rate = 120
The average delay = 0.00117551 +- 8.95762e-05
The utilization = 0.618032 +- 0.01041
The throughput = 1211.83
The collision rate = 228.763

For an arrival rate = 140
The average delay = 0.00229173 +- 0.000480587
The utilization = 0.718528 +- 0.0163306
The throughput = 1408.88
The collision rate = 648.538
```

```
For an arrival rate = 160
The average delay = 0.0147868 +- 0.00969014
The utilization = 0.79573 +- 0.00643513
The throughput = 1560.25
The collision rate = 2016.56

For an arrival rate = 180
The average delay = 0.0859435 +- 0.00865851
The utilization = 0.780971 +- 0.00143146
The throughput = 1531.32
The collision rate = 2860.98
```

These results are summarized in Table 6.2.

Table 6.2 Simulation results.

Arrival Rate	Arrival Delay	Utilization	Through-put	Collision Rate
20.0	0.000570 ± 4.44E−06	0.102 ± 0.00292	200.47	0.667
40.0	0.000616 ± 9.42E−06	0.202 ± 0.00448	396.89	5.065
60.0	0.000670 ± 1.16E−05	0.306 ± 0.00749	599.80	15.65
80.0	0.000782 ± 1.57E−05	0.419 ± 0.00886	820.97	52.11
100.0	0.000936 ± 4.76E−05	0.517 ± 0.00857	1013.5	115.34
120.0	0.001175 ± 8.96E−05	0.618 ± 0.01412	1211.8	228.76
140.0	0.002291 ± 0.000480	0.718 ± 0.01633	1408.88	648.5
160.0	0.014786 ± 0.009690	0.796 ± 0.00643	1560.2	2016.5
180.0	0.085943 ± 0.008658	0.781 ± 0.00143	1531.3	2860.9

6.4 Summary

In this chapter, we have discussed simulation of CSMA/CD-based local area networks. After a brief description of CSMA/CD LANs, we have presented a step-by-step process of simulating this type of LAN. A simulation program is also described in detail, from providing input parameters to printing output results. The output results are presented in terms of the average delay per packet, the average throughput, the average utilization, and the average collision rate. The results are presented with 95% confidence intervals.

References

[1] J. L. Hammond and P. J. O'Reilly, *Performance Analysis of Local Computer Networks*, Reading, MA: Addison-Wesley, 1986.

[2] M. Ilyas and H. T. Mouftah, *Simulation Tools for Computer Communication Networks*, Conference Record of the IEEE GLOBECOM '88, November-December 1988, pp. 1702–1706.

Problems

1. Modify the simulation program to find the average duration of the busy (including both successful and unsuccessful) period on the transmission medium.
2. Modify the simulation program to find the average packet delay for each station. Compare these delays to see if the access mechanism introduces any positional unfairness.
3. The amount of time a LAN spends in activities other than actual transmission of user information is called overhead. Modify the simulation programs to compute the overhead time.
4. Modify the simulation program to obtain the delay distribution.

2. [illegible]
3. [illegible]

Chapter 7

Simulation of Star LANs

There is no man living who cannot do more than he thinks he can.

—Henry Ford

7.1 Introduction

In this chapter, we discuss simulation of star local area networks (LANs). These LANs have traditionally been used for circuit-switched application, and their operation is very similar to that of polling systems [1]. The most important aspect of star LANs that makes them different from other LANs is their centralized control, which is responsible for communication among all stations connected to the network. Unfortunately, this aspect also makes star LANs less popular because of the vulnerability of the central controller.

Section 7.2 contains a brief description about the operation of star local area networks. In Section 7.3, we describe the structure of the simulation model presented in this chapter. The same section contains assumptions made in the simulation process. A few typical simulation results are also presented. Section 7.4 summarizes the contents of this chapter.

7.2 Operation of Star LANs

Like all local area networks, star LANs provide communication among network users/stations through a shared set of facilities. A star LAN consists of a central controller and transmission lines connecting stations to the controller. The central controller establishes the connection, facilitates the communication, and terminates the connection at the end of the communication among any pair of network stations. A typical star LAN is shown in Figure 7.1. Because of the level of responsibility associated with the single central controller in star LANs, their reliability has always been questionable. This is one of the reasons why these networks do not enjoy as much popularity as token-passing LANs or CSMA/CD-based LANs.

The operation of a star LAN is very similar to that of polling systems [1]. A station waits until it is asked by the central controller whether or not it has some information to transmit. If the station has some information to transmit, it says so, and the central controller establishes the connection between the source station and the destination station. The source station then transmits its information packets for no more than a specified duration of time. The central controller terminates the connection at the end of communication and directs its question to the next station. If the station being asked does not have any information to transmit, it sends a negative reply to the central controller, and the controller repeats the question to the next station. This process continues in a cyclic fashion.

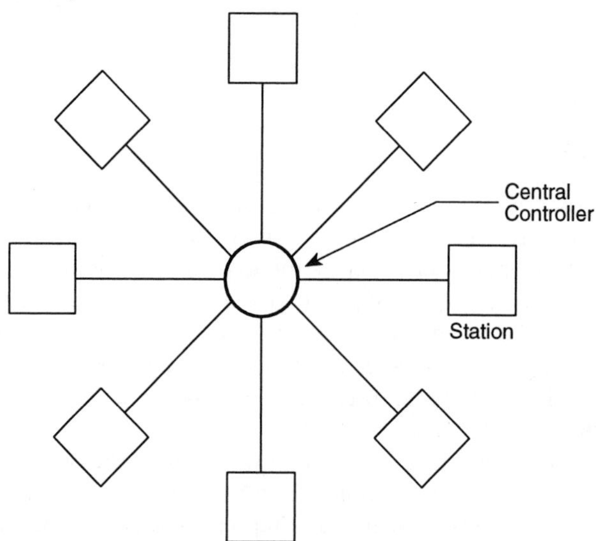

Figure 7.1 A typical star local area network.

7.3 Simulation Model

This section describes the simulation model developed for simulating local area star networks. The program uses the event-scheduling approach, and the simulation process is self-driven [2, 3]. The program is written in a turbo C. Generation of random numbers according to a desired probability distribution is a necessity for any self-driven simulation. In this simulation program, we have used a built-in function `rand()` for this purpose. This function generates uniformly distributed random numbers, which are then converted to follow a desired probability distribution. In this simulation program, we primarily need exponentially distributed random numbers, and the output of the `rand()` function can be easily converted to exponentially distributed numbers using the transform method [3].

A flowchart of the simulation program that shows the structure of the program and the logical steps involved is given in Figure 7.2. The simulation program has four major sections: initialization, processing, control, and output. In the initialization section, values of the input parameters are read and all the variables are initialized to their appropriate values. This includes initialization of the event list that contains the timing at which various events are supposed to take place. This section prepares the simulation for executing events.

The next logical step is to scan the event list and pick an event with the shortest time of occurrence. This process also identifies the event as an arrival of a packet, a departure of a packet, or an arrival of a polling packet at a station. The program then executes the selected event and updates the values of all the variables affected by the event. This is part of the processing section. After an event has been successfully completed, the program's control section checks to see if the simulation should be stopped. If the decision is to continue the simulation, then the event list is scanned again to pick the next event. This process continues until enough packets have been transmitted and delivered to their respective destinations to yield reasonably converged simulation results. If the decision is to stop the simulation, the output section computes the final simulation results and prints them out, and the simulation process stops.

In the next section, we describe the assumptions used in the simulation process. Section 7.3.2 describes the input and output variables used in this program. In Section 7.3.3, the simulation program is described in detail, and in Section 7.3.4, some typical simulation results are presented.

7.3.1 Assumptions

The following assumptions have been made in the simulation process:

- Arrivals at all stations follow a Poisson process.
- All stations generate the traffic at the same rate.
- Packet lengths are fixed.
- The transmission medium is assumed to be error-free.
- Propagation delay is 5 microseconds per kilometer of transmission medium.

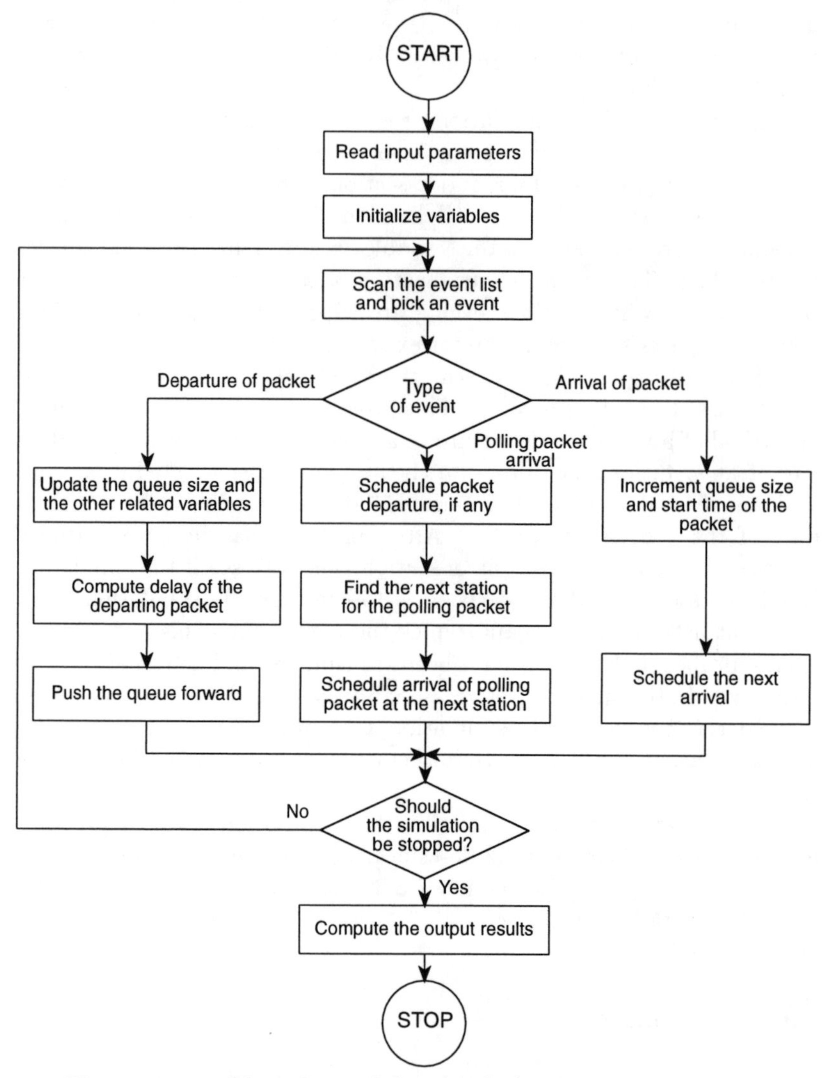

Figure 7.2 Flow chart of the simulation program.

7.3.2 Input and Output Variables

Input Variables:

`MAX_STATIONS` Number of stations in the local area network.

`PACKET_LENGTH` Packet length in bits. Its reasonable value is from 500 bits to 10,000 bits.

`POLLING_PKT_LENGTH` Polling packet length in bits. Its reasonable value is about 50 bits.

`BUS_LENGTH` Length of the transmission link in meters from each station to the central hub. Propagation delay is calculated assuming that it is 5 microseconds per kilometer. Its reasonable value is 1 km to 5 km.

`RATE` Transmission rate of the transmission medium in bits per second. Its reasonable value is 1.0 Mbps to 5.0 Mbps.

`MAX_PACKETS` The number of packets for which each simulation run is executed before terminating it. At lower traffic loads a value of about 5,000 can be used. However,at higher loads a value of at least 20,000 should be used. Also, if the size of the network grows, this value should be increased, too. Basically this parameter is for convergence of simulation results.

`FACTOR` An accuracy parameter that determines the unit of time for a simulation run. At higher transmission rates this parameter should have a higher value for reasons of accuracy. If `FACTOR` = 1.0, the time unit is seconds. If `FACTOR` = 1,000, the time unit is milliseconds, and so on. For our simulation, the value of `FACTOR` should be at least 1,000.

`MAX_Q_SIZE` The maximum buffer size at a station.

`DEGREES_FR` The degrees of freedom to be used for calculating confidence intervals for the output results. In our simulation the range for this variable is from 1 to 10. However, in most of the results, we have used 5 degrees of freedom and a 95% level of confidence.

Output Variables:

`average_delay` The average delay per packet in seconds per packet.

`delay_con_int` The 95% confidence interval for the average delay with the selected degrees of freedom.

7.3.3 Description of the Simulation Model

In this Section we describe the simulation program step by step and explain how various aspects of a star local area network are simulated. A complete listing of the program is given in Appendix C.

The following segment of the simulation program includes declaration statements indicating constants, real variables, and integer variables. The segment also includes the header files for various built-in functions of the turbo C language used in the simulation process. We also assign appropriate values to the input parameters for a simulation run.

```
# include    <stdio.h>
# include    <stdlib.h>
# include    <math.h>


# define   MAX_STATIONS           50           /* Number of stations */
# define   RATE                   5000000.0    /* Transmission rate in
                                               bits per second */
# define   PACKET_LENGTH          1000.0       /* Packet length (bits) */
# define   POLLING_PKT_LENGTH     10.0         /* Poll packet length (bits) */
# define   BUS_LENGTH             1000.0       /* Bus length in meters */
# define   MAX_PACKETS            1000         /* Maximum packets to be
                                               transmitted in simulation run */
# define   FACTOR                 1000.0       /* A factor used for
                                               changing units of time */
# define   MAX_Q_SIZE             100          /* Maximum queue size */
# define   DEGREES_FR             5            /* Degrees of freedom */

float arrival_rate; /* Arrival rate (in packets/sec) for each station */
float tau; /* End-to-end propagation delay */
float packet_time; /* Packet transmission time */
float polling_pkt_time; /* Polling packet transmission time */
float start_time[MAX_STATIONS][MAX_Q_SIZE]; /* Starting time of packets */
float event_time[MAX_STATIONS][3]; /* Time of occurrence of an event */
float t_dist_par[10] = {12.706, 4.303, 3.182, 2.776, 2.571, 2.447, 2.365,
                  2.306, 2.262, 2.228}; /* T- distribution parameters */
float delay_ci[DEGREES_FR + 1]; /* An array to store delay values */


float rho, clock, no_pkts_departed, next_event_time;
float x, logx, rand_size, infinite;
float delay, total_delay, average_delay;
float delay_sum, delay_sqr, delay_var, delay_sdv, delay_con_int;


int queue_size [MAX_STATIONS]; /* Current queue size at a station */
int i, j, ic, ii, next, next_station, next_event;
```

The array variables in these statements are explained next.

`start_time[i][j]` The starting time of a packet that is sitting in the buffer of the ith station and occupies the jth position; used in calculating the delay when the packet departs from the station.

`event_time[i][j]` The time at which an event of type j is to occur at the ith station; $j = 0$ implies a packet arrival event, $j = 1$ implies a packet departure event, and $j = 2$ implies a polling packet arrival.

`queue_size[i]` Queue size or buffer occupancy at the ith station.

`t_dist_par[i]` This array contains T-distribution parameters used in calculating confidence intervals. The ith element of this array indicates the parameters to be used in calculating 95% confidence intervals with i degrees of freedom.

`delay_ci[i]` This array is used to temporarily hold the values of delay from various simulation runs with the same input parameters. These values are eventually used in calculating confidence intervals.

The main body of the program starts from the next segment. The segment starts with some output statements. The variable `arrival_rate` is initialized to zero, and its values are incremented with a do-loop. Some of the input variables that are assigned one type of units in the beginning are converted to a convenient type of units in this segment. This is done merely for programming convenience. For example, the values of PACKET_LENGTH and POLLING_PKT_LENGTH are initially assigned values in bits and are then converted to their equivalent time units (i.e., the time it takes to transmit that many bits). The corresponding variables in time units are packet_time and polling_pkt_time, respectively. The variable tau represents the end-to-end propagation delay, and its value is calculated assuming that the wave propagation speed on the transmission medium is 5 microseconds per kilometer.

```
main ()
{
printf("The following results are for:  \n");
printf("Degrees of freedom = %d\n", DEGREES_FR);
printf("Confidence interval = 95 percent \n");
printf(" ======================================== \n");
printf("\n");

arrival_rate = 0.0;
tau = BUS_LENGTH * FACTOR * 5.0 * pow (10.0, -9.0);
packet_time = PACKET_LENGTH * FACTOR / RATE;
polling_pkt_time = POLLING_PKT_LENGTH * FACTOR / RATE;
```

The following two do-loops are used to run simulation for various values of arrival rates and to conduct several simulation runs (with the same values of input variables but with different random number streams) for the purpose of calculating confidence intervals.

```
for (ii=0; ii < 10; ii++)
        {
        arrival_rate += 20.0;
        for (ic = 0; ic <= DEGREES_FR; ic++)
         {
```

In the next segment of the program, we initialize all variables to their appropriate values. The clock for the simulation process is represented by clock and is initialized to 0.0. The variable no_pkts_departed represents the total number of packets delivered to their destinations and is initialized to 0.0. The variables total_delay and average_delay are also initialized to 0.0. The variable rand_size represents the largest integer value that the program can handle. This depends on the computer being used. Thus, in order to have flexibility of running the simulation program on any computer, the program uses the sizeof(int) function of the turbo C language to determine the size of the integer in bytes. That information is used to determine the value of variable rand_size, which is then used in random number generation.

```
rho                  = 0.0;
clock                = 0.0;
no_pkts_departed     = 0.0;
total_delay          = 0.0;
next_event_time      = 0.0;
average_delay        = 0.0;
infinite             = 1.0 * pow (10.0, 30.0);
rand_size            = 0.5 * pow (2.0, 8.0 * (float) sizeof(int));
```

The next step is to see if the traffic load is too much for the network to carry. We calculate the traffic intensity rho and check to see if it exceeds the network capacity. If it does, the network is assumed to be overloaded, we terminate the simulation prematurely and notify the user. If it does not exceed the network capacity, the program proceeds to the next step, initializing variables.

```
rho = arrival_rate * MAX_STATIONS / RATE;
if (rho >= 1.0)
        {
        printf("Traffic intensity is too high");
        exit(1);
        }
```

In the next do-loop, we initialize the variable queue_size[i] to zero for all stations and we also initialize the variable start_time[i][j] to zero for all possible entries.

```
for (i = 0; i < MAX_STATIONS; i++) queue_size[i]=0;
for (i = 0; i < MAX_STATIONS; i++)
        {
        for (j = 0; j < MAX_Q_SIZE; j++) start_time[i][j]=0.0;
        }
```

The following nested do-loop initializes the event list. All events except the packet arrival events are disabled. This is done because no other event can take place before packets start arriving at stations. In order to disable an event, a very large value (of the order of 1.0E+30) can be assigned to the event. When the event list is scanned, the event with the smallest value is selected first. By doing so we force the packet arrivals to take place first, and then we schedule subsequent events. The polling packet arrival event is also enabled initially at one of the stations so as to begin the polling process.

```
for (i = 0; i < MAX_STATIONS; i++)
        {
        for (j = 0; j < 3; j++)
         {
         event_time[i][j] = 0.0;
         if (j !=0) event_time[i][j] = infinite;
         if (i==0 && j==2) event_time[i][j] = 0.0;
         }
        }
```

After all the variables have been initialized, the next segment of the program scans the event list and picks the next event to be processed. The scanning process is merely a process of picking an event with the smallest time value associated with it. This produces three values: next_station, next_event and next_event_time. The variable next_event_time represents the time of occurrence of the selected event, and that is why it is assigned a very large value before the scanning process begins. The variable next_station indicates the station at which the selected event is to take place, and next_event denotes the type of the selected event. The value of **next_event** determines which section of the program should execute the selected event.

The process of scanning the event list, picking the next event, and executing it continues until a sufficient number of packets have been handled by the network. This is the job of the control section of the program. The first statement of the following segment represents the control section. This segment checks if a desired number of packets have departed so that the simulation may be stopped. Specifically, it checks to see if the total number of packets delivered (**no_pkts_departed**) is less than the desired maximum (**MAX_PACKETS**). If so, it is the program continues the scanning of the event list and picks the next event to process. If not, the program will go to the output section of the program.

As the variable **next_event_time** indicates the time of the selected event, the value of **clock** is immediately equated to that of **next_event _time** as shown below.

```
while (no_pkts_departed < MAX_PACKETS)
        {
        next_event_time = infinite;
        for (i = 0; i < MAX_STATIONS; i++)
         {
         for (j = 0; j < 3; j++)
          {
          if (next_event_time > event_time[i][j])
           {
           next_event_time = event_time[i][j];
           next_station = i;
           next_event = j;
           }
          }
         }
        clock = next_event_time;
```

After scanning the event list, if the program does not find a legitimate event (packet arrival, packet departure, or polling packet arrival) to process, it will execute the following short segment. This segment notifies the user that there is some problem with the event list and stops the program. These are some debugging aids that have been built into the program for helping the new users.

```
if (next_event > 2)
        {
        printf("Check the event list");
        exit(1);
        }
```

If the selected event is a legitimate event, then an appropriate segment is chosen with the help of the following switch statement.

```
switch (next_event)
```

If the selected event happens to be an arrival of a packet (**next_event** = 0), the following segment of the program updates the values all the affected variables in the program and schedules the next packet arrival before scanning the event list for the next event.

The first thing to do is to see if the selected event is an arrival of packet. If it is (**next_event** = 0), this segment is executed; otherwise the program checks to see if some other event has been selected. When a packet arrival occurs, the queue size (**queue_size**) at the selected station (**next_station**) is incremented by one. If this arrival has caused the queue size to exceed its limit, the program notifies the user and stops. The start time (**start_time**) of the newly arrived packet is also initialized to the current simulation clock value, and the packet is placed at the appropriate place in the queue.

Before leaving this segment, the next arrival at the same station is scheduled. Packet arrivals are assumed to follow a Poisson process, which means that the packet interarrival times are exponentially distributed. In order to determine when the next arrival will take place, we need exponentially distributed random numbers. We use the **rand()** function for generating uniformly distributed numbers between 0 and 1 and then convert them to exponentially distributed random numbers using the transform method. After scheduling the next arrival event, we go to that part of the program, where we check to see if the simulation should be terminated.

```
{
case 0:  /* This is an arrival event.  */
        {
        queue_size[next_station] ++ ;
        if (queue_size[next_station] > MAX_Q_SIZE)
         {
         printf("The queue size is large and is = %d\n",
         queue_size[next_station]);
         exit(1);
         }
        start_time [next_station][(queue_size[next_station]-1)] = clock;

    /* Schedule the next arrival */

    for (;;)
     {
     x = (float) rand();
     if(x != 0.0) break;
     }
    logx = -log(x/rand_size) * FACTOR / arrival_rate;
    event_time[next_station][next_event] = clock + logx;
    break;
    }
```

If the event selected after the event list is scanned happens to be a departure event (**next_event** = 1), the following segment will update all the affected variables. In this segment, we decrement the queue size, calculate the delay for the departing packet, and update the values of other affected variables.

First of all we check and make sure that the selected event is a departure event before updating the variables. If the selected event is a departure event (**next_event** = 1), processing will start by decrementing the queue size (queue_size) at the station by 1. Then the total number of packets departed (no_pkts_departed) is incremented by 1, the packet delay is computed by subtracting the packet starting time (start_time) from the current value of the simulation clock, and the total delay of all packets departed so far is updated (for averaging purpose).

```
case 1:  /* This is a departure event.  */
         {
   queue_size[next_station] -- ;
   no_pkts_departed ++ ;
   delay = clock - start_time[next_station][0];
   total_delay += delay;
```

Once a packet has departed from a station's queue, the remaining packets must be pushed forward. This is done only if the queue size is more than zero. As the packets move forward in a queue, they carry their start times with them. The start time of the departing packet is reset to zero. The process of pushing the queue forward at the selected station is done in the following segment of the program. As we allow only one departure at a time, we disable departures at the current station before leaving this segment.

```
/* Push the queue forward.  */

if (queue_size[next_station] > 0)
         {
         for (i=0; i < queue_size[next_station]; i++)
          start_time[next_station][i] = start_time[next_station][i+1];
          }
start_time[next_station][queue_size[next_station]]=0.0;
event_time[next_station][next_event] = infinite;
break;
}
```

The following segment of the program updates all the affected variables if the selected event happens to be an arrival of the polling packet (next_event = 2). The first item (before starting the processing of this segment) ensures that the event selected is the arrival of the polling packet. Then this segment starts with disabling the polling packet arrival event at the selected station (next_station) and the next station to receive the polling packet. It then checks the queue size at the selected station to see if there are any information packets to be transmitted. If there are some packets, we schedule the departure of a packet. We also schedule the arrival of a polling packet at the next station. However, if there are no information packets to be transmitted at the selected station, we disable the departure event at that station and schedule the arrival of the polling packet at the next station.

```
case 2:  /* Find the next station to be serviced.  */
         {
         event_time[next_station][next_event] = infinite;
         next = next_station + 1;
         if(next == MAX_STATIONS) next = 0;
         if (queue_size[next_station] > 0)
          {
          event_time[next_station][1] = clock + packet_time + tau;
```

```
                event_time[next][2] = clock + packet_time + polling_pkt_time
                + tau;
                }
               else
                {
                event_time[next_station][1] = infinite;
                event_time[next][2] = clock + polling_pkt_time + tau;
                }
               break;
               }
              }
    }
```

Once it has been decided that the simulation process is to be stopped, the following segment of the simulation program computes the average delay per packet for this simulation run. It is computed by the total delay of all packets (**total_delay**) divided by the total number of packets delivered (**no_pkts_departed**). It is also divided by **FACTOR** to convert the average delay in units of seconds per packet. The segment stores the values of delay in an array (**delay_ci**) until a specified number of simulation runs have been made for the calculation of confidence interval for the average delay per packet.

```
    average_delay = total_delay / (no_pkts_departed * FACTOR);
    delay_ci[ic] = average_delay;
    }
```

Once all the simulation runs have been made, the following segment computes the average delay (**delay_sum**) from all the simulation runs and the 95% confidence interval (**delay_con_int**) for the average delay. The output results are then printed before the simulation program stops.

```
    delay_sum = 0.0;
    delay_sqr = 0.0;
    for (ic = 0; ic <= DEGREES_FR; ic++)
             {
             delay_sum += delay_ci[ic];
             delay_sqr += pow(delay_ci[ic],2.0);
             }
             delay_sum = delay_sum / (DEGREES_FR + 1);
             delay_sqr = delay_sqr / (DEGREES_FR + 1);
             delay_var = delay_sqr - pow(delay_sum,2.0);
             delay_sdv = sqrt(delay_var);
             delay_con_int = delay_sdv * t_dist_par[DEGREES_FR-1] /
             sqrt(DEGREES_FR);
             printf("For an arrival rate = %g\n",arrival_rate);
             printf("The average delay = %g", delay_sum);
             printf(" +- %g\n", delay_con_int);
             printf("\n");
             }
    }
```

Another do-loop increments the values of **arrival_rate** (by 20.0) and repeats the simulation process for a specified number of times. There are several more output results that can be easily computed from this program [3]. Some of them are delay histograms, the average queue size, and so on.

7.3.4 Typical Simulation Sessions

In this section, we present a typical simulation run for star local area networks. The input values are assigned as constants at the beginning of the program. The arrival rate per station is adjusted automatically within the simulation. After the values of the input parameters have been defined as constants, the program is compiled and executed by giving an appropriate command for the computer in use.

For the purpose of illustration, the following values are assigned to the input parameters:

```
MAX_STATIONS = 50
RATE = 5000000.0
PACKET_LENGTH = 1000.0
POLLING_PKT_LENGTH = 50.0
BUS_LENGTH = 1000.0
MAX_PACKETS = 1000
FACTOR = 1000.0
MAX_Q_SIZE = 100
DEGREES_FR = 5
```

The output results of the simulation program for the given input parameters are as follows:

```
The following results are for:
Degrees of freedom = 5
Confidence interval = 95 percent
========================================

For an arrival rate = 20
The average delay = 0.000995696 +- 2.51265e-05


For an arrival rate = 40
The average delay = 0.00136352 +- 7.37867e-05


For an arrival rate = 60
The average delay = 0.00214825 +- 0.000160036


For an arrival rate = 80
The average delay = 0.00463656 +- 0.00128993


For an arrival rate = 100
The average delay = 0.0183729 +- 0.00409046


For an arrival rate = 120
The average delay = 0.03226 +- 0.00314482


For an arrival rate = 140
The average delay = 0.041015 +- 0.00233979


For an arrival rate = 160
The average delay = 0.0483129 +- 0.00235372


For an arrival rate = 180
The average delay = 0.0546609 +- 0.00236092
```

```
For an arrival rate = 200
The average delay = 0.0599928 +- 0.00235157
```

These results are presented in Table 7.1.

Table 7.1 Simulation results for a star LAN.

Arrival Rate (in packets per second per station)	Average Delay per Packet (in seconds)
20.0	0.000995696 ± 2.51265E−05
40.0	0.00136352 ± 7.37867E−05
60.0	0.00214825 ± 0.000160036
80.0	0.00463656 ± 0.00128993
100.0	0.0183729 ± 0.00409046
120.0	0.03226 ± 0.00314482
140.0	0.041015 ± 0.00233979
160.0	0.0483129 ± 0.00235372
180.0	0.0546609 ± 0.00236092
200.0	0.0599928 ± 0.00235157

7.4 Summary

In this chapter we have discussed simulation of star local area networks. After a brief description of the operation of star LANs, we have explained a simulation program step by step, from providing input parameters to printing output results. The output results are presented in terms of the average delay per packet and 95% confidence intervals.

References

[1] J. L. Hammond and J. P. O'Reilly, *Performance Analysis of Local Computer Networks.* Reading, MA: Addison-Wesley 1986.

[2] M. Ilyas and H. T. Mouftah, *Simulation Tools for Computer Communication Networks*, Conference Record of the IEEE GLOBECOM '88, Nov./Dec. 1988, pp. 1702–1706.

[3] R. Jain, *The Art of Computer Systems Performance Analysis.* New York: John Wiley and Sons, 1991.

Problems

1. Modify the simulation program to find out the average number of token circulations per unit time in star LANs.
2. Modify the simulation program to find out the average number of stations that transmit in one token circulation.
3. The amount of time a LAN spends in activities other than actual transmission of user information is called overhead. Modify the simulation programs to compute the overhead time in star LANs.
4. Modify the simulation program to obtain the delay distribution.

Chapter 8

Simulation Languages

See everything ..., overlook a great deal..., correct a little
— Pope John XXIII

8.1 Introduction

The purpose of this chapter is to present the characteristics of common simulation languages and provide the analyst with the criteria for choosing a suitable language. Once the analyst has acquired a thorough understanding of the system to be simulated and is able to describe precisely how the model would operate, the next step is to decide on the language to use in the simulation. This step should not be taken lightly, for the choice of a language has several implications, some of which will be discussed later. After deciding on the language to apply, the analyst needs to consult the language reference manual for all the details.

There are basically two types of languages used in simulation: multipurpose languages and special-purpose languages. The former are *compiler languages*; the latter are *interpretive languages* [1]. A compiler language is made up of macrostatements and requires compilation and assembly before execution can occur. An interpretive language consists of symbols that denote commands to carry out operations directly without the need for compilation and assembly. Thus, the major difference between the two types of languages is the distinction between a *source* program and an *object* program. An analyst usually submits a source program to a computer. If the source program is in a compiler language, an object program is needed for execution. If the source program is in an interpretive language, execution is done directly without any object program.

Some analysts tend to select multipurpose or general-purpose languages such as FORTRAN, BASIC, PASCAL, and C for the simulation of local area networks. Although these languages are far from ideal for discrete simulation, they are widely used. Why? There are at least three reasons. First, there is conservatism on the part of the analysts and organizations that support them. Many organizations are committed to multipurpose languages and do not want to be vulnerable to a situation where a code written in a language only familiar to an analyst may have to be rewritten when the analyst leaves the organization. Second, the widespread availability of multipurpose languages and the libraries of routines that have been developed over the years makes them more desirable. It is easy to gain technical support because experts in multipurpose languages are everywhere. Third, highspeed in the simulation is possible if a general-purpose language is used. Analysts who prefer fast-running simulations use a general-purpose language. In view of the problem of learning another set of syntactic rules, a decision in favor of a general-purpose language is often considered wise by analysts.

The development of special-purpose simulation languages began in the late 1950s. The need came from the fact that many simulation projects required similar functions across various applications. The purpose of simulation languages is to provide the analyst with a relatively simple means of modeling systems. Unlike general-purpose languages such as FORTRAN, where the analyst is responsible for all the details in the model, special-purpose languages are meant to eliminate the major portion of the programming effort by providing a simulation-oriented framework about which a model is constructed in a simple fashion. Although many such languages have been developed, only a few have gained wide acceptance.

Before deciding which type of language to use in a simulation, the analyst must carefully weigh the advantages of the multipurpose languages against the almost guaranteed longer program development and debugging time required in special-purpose languages. Irrespective of the language used in the simulation of a LAN, the language must be capable of performing functions including

- Generating random numbers
- Executing events
- Managing queues
- Collecting and analyzing data
- Updating simulation time

This chapter introduces and provides basic insight into the basic features of six of the more commonly used special-purpose, discrete simulation languages. The languages we consider are GPSS, SIMSCRIPT, GASP, SIMULA, SLAM, and RESQ. No attempt will be made to include the many instructions available in these languages. Our objective is only to introduce the basic instructions and provide a flavor of the languages. Interested readers must consult the manuals and references for more details.

8.2 GPSS

The most commonly used special-purpose simulation language is GPSS (General Purpose Simulation System). It is an interpretive, dis- crete-event simulation language originally developed by Geoffrey Gordon at IBM and first published by him [2] in 1961. The language has gone through several revisions to the point where there are now two versions: GPSS/360 and GPSS V. When IBM stopped enhancing GPSS in 1972, other vendors improved GPSS. Such improved versions include GPSS/H, developed by James Henriksen in 1977 as a compiler language [3], and GPSS/PC, developed in 1988 for the IBM PC and compatibles [4]. Another version is GPSS-FORTRAN Version 3, which is a simulation package consisting of a main program in FORTRAN 77 and over a hundred subroutines [5].

The language consists of powerful commands that relieve the analyst of the need to code many special simulation features. It is basically a flow-oriented or block-diagram-oriented language. It requires that the analyst construct a flowchart from which the simulation program is developed. This is easily achieved because the modeling blocks are also the basic programming statements. This feature makes GPSS attractive for many queueing problems. The language is easy to learn and does not require prior programming experience.

GPSS is based on events called *transaction blocks*, with each block operating on entities that flow from one block to another. It has some 50 different blocks for constructing flow charts and the analyst is restricted to these. A model is constructed by connecting a certain number of blocks together. Entities are created by a **GENERATE** transaction block and are destroyed by a **TERMINATE** transaction block. The **GENERATE** block essentially generates interarrival times. For example, with the statement

```
GENERATE A,B
```

entities are generated every A $\pm$ B time units. That is, uniformly random time spaces between entities of A $\pm$ B time units separation are produced. The **TERMINATE** statement serves to remove entities from the model or to terminate the simulation run. For example, with the statement

```
TERMINATE A
```

an entity is removed from the model and the start count is decremented by A. The **START** block is used to specify the length of the simulation run. Its operand is determined in conjunction with the **TERMINATE** operands in the various segments of the program. The **ADVANCE** block simulates the passage of time while an entity is in a given state, such as when a customer is waiting in a queue. For example, with the statement

```
ADVANCE A,B
```

the entity remains in the **ADVANCE** statement block for A $\pm$ B time units, uniformly distributed. The **SEIZE** and **RELEASE** blocks correspond to the transaction of seizing a facility and releasing it after the transaction has been completed. They control entry and exit from the **ADVANCE** statement. The sequence of blocks is as follows:

```
SEIZE A
ADVANCE B
RELEASE A
```

which means that facility A is used for B time units. The **QUEUE** and **DEPART** blocks are included in a GPSS program to gather statistics. For example,

```
QUEUE A
  :
DEPART A
```

will cause the statistics (number of entries, average time spent, etc.) at queue A to be automatically printed out at the end of the simulation. The **SIMULATE** block indicates the start of the program. It is placed at the top of the program and commands that the simulation be performed. The **END** block denotes the end of the program. It is included at the end of the program to indicate that a program is completed.

In GPSS, a comment line is indicated by an asterisk (*) at the beginning of the line. A comment can also be appended to a GPSS statement by leaving a blank between the final argument of the statement and the comment. For a more complete description of GPSS blocks, one should consult the literature [2, 6–8].

With the few GPSS statements that we have covered, a simple model may now be implemented for the purpose of illustration. Figure 8.1 shows the flowchart using the standard GPSS symbols for blocks, and Figure 8.2 presents the program listing for the simulation of the M/M/1 queue in GPSS. In Figure 8.2, the line numbers are not part of the program but are inserted for pedagogic reasons. Line 6 in the program shows generation of customers with exponential interarrival times having mean 1.0 and using random number stream 1. Other lines are self-explanatory.

The advantages of GPSS are as follows:

1. It is flexible and easy to learn. Coding from the flowchart or block diagram is straightforward.
2. It handles all the details of the basic simulation functions. Its built-in features can drastically reduce the amount of time that must be devoted to coding and debugging a simulation model.
3. Many statistics of interest are printed automatically.

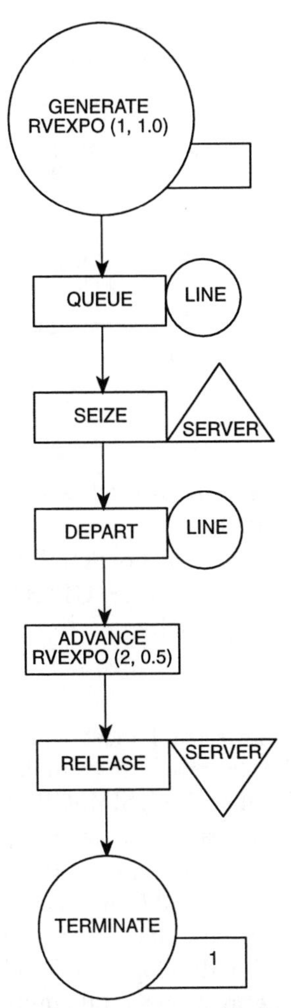

Figure 8.1 GPSS flowchart for the M/M/1 queue.

GPSS has some shortcomings, which include the following:

1. It does not allow subprogramming and other usual features.
2. The arithmetic capabilities are poor: Complex computation can be very difficult to execute. Extensions to the statistics automatically gathered by GPSS may require rather tortuous programming.
3. It does not have a built-in random variate generator and must resort to some form of interpolation to generate random variates.
4. All clock times in the simulator are integer values 1, 2, 3, ... This forces the analyst to select a GPSS time unit to obtain the desired precision. The absence of a floating-point capability in GPSS is an unfortunate shortcoming and can lead to serious errors.

```
1  **********************************************
2  *  PROGRAM TO SIMULATE THE M/M/1 QUEUE
3  *
4  ***********************************************
5   SIMULATE
6   GENERATE  RVEXPO  (1, 1.0)     CREATE ARRIVING CUSTOMERS
7   QUEUE     LINE                 ENTER THE QUEUE
8   SEIZE     SERVER               BEGIN SERVICE
9   DEPART    LINE                 EXIT THE QUEUE
10  ADVANCE   RVEXPO  (2, 0.5)     DELAY FOR SERVICE
11  RELEASE   SERVER               FINISH SERVICE
12  TERMINATE 1                    REMOVE CUSTOMERS FROM SYSTEM
13 *
14  START     2000                 PROCESS 2000 CUSTOMERS
15  END
```

Figure 8.2 GPSS program listing for the simulation of the M/M/1 queue.

5. Most simulation models written in GPSS execute more slowly and hence are more expensive than models written in general-purpose languages such as FORTRAN.
6. GPSS is not as widely available as FORTRAN.

In spite of these disadvantages, however, GPSS is a widely used, powerful, and well-known simulation tool.

8.3 SIMSCRIPT

SIMSCRIPT is perhaps the second most widely used special-purpose programming language for discrete-event simulation. The language has its roots in FORTRAN. Unlike GPSS, SIMSCRIPT approaches a general-purpose language and can be learned as the first programming language.

SIMSCRIPT was originally developed at Rand Corporation in the early 1960s by Markowitz, Karr, and Hausner [9]. The language has been modified and improved over the years. Some people regard the more recent version SIMSCRIPT II.5, developed at Rand by Kiviat, Villanueva, and Markowitz [10], as the most powerful simulation language used in computer modeling. Its English-like instructions make SIMSCRIPT II.5 self-explanatory and self- documenting. It not only contains the capabilities for building discrete-event, continuous, or combined simulation models, it is often used for large, complex simulation models, especially when the system is not queueing oriented. Our discussion here will be on this version of SIMSCRIPT.

SIMSCRIPT II.5 is divided into five separate levels:

Level 1: A simple language designed to introduce programming to beginners. It resembles simplified FORTRAN or BASIC.

Level 2: A language comparable to FORTRAN. It has all the I/O formatting of FORTRAN. The limited data structures of FORTRAN have been improved in this level.

Level 3: A language comparable to ALOG or PL/I. The I/O facilities are more powerful and recursive subroutines are allowed.

Level 4: A refinement of data structures and algorithms central to the discrete- event technique.
Level 5: A self-contained simulation language. It contains statements for time advance, event processing, random variate generators, automatic statistics gathering, and diagnostic facilities.

An analyst using SIMSCRIPT employs three principal building blocks: *entities, attributes,* and *sets.* An entity is a structured variable, an object of the system. The attributes of an entity are its characteristics or fields within its structure. Sets denote entity groupings and relationships.

A simulation program in SIMSCRIPT is usually composed of three parts: the preamble, the main program, and the event routines. As shown in Figure 8.3, each part begins with a keyword and ends with an `END` statement. Every simulation program must contain only one preamble, one main program, and one or more event routines. It may or may not have subprograms, which are similar to subroutines in FORTRAN. The preamble section declares or defines the system's entities, attributes, and sets. All global variables and the statistics to be collected are defined here. The main program section is where program execution begins. Initiation is followed by scheduling of the first event. The control is passed to a built-in timing routine by executing the `START SIMULATION` statement. The event routine section further describes each process defined in the preamble and the necessary action to be taken when the events occur. Typical SIMSCRIPT declarations, executive statements, I/O statements, and entity statements are shown in Table 8.1. Every statement begins in SIMSCRIPT with a keyword; all system defined variables end with a period followed by a letter. For example,

`TIME.V` (current simulation time—variable)
`EXP.C` (2.7182818—constant)
`SIN.F` (sine of expression—function)
`GABB.R` (user-supplied routine called by SIMSCRIPT—routine)

The mode of variables may be defined as `INTEGER`, `REAL`, or `ALPHA` in the preamble; the default option is `REAL`. SIMSCRIPT has some built-in functions such as `ABS.F`, `COS.F`, `RANDOM.F`, `POISSON.F`, `EXPONENTIAL.F`, and `BETA.F`.

With this brief introduction, a SIMSCRIPT model of the M/M/1 queue will now be considered as an illustration. The example is a slight modification of the one described in Law and Kelton [12]. The model is shown in Figure 8.4(a), and its output is in Figure 8.4(b). The `TALLY` statement in the preamble requests certain statistics to be collected. We notice that unlike GPSS, the simulation results are not automatically collected by SIMSCRIPT. The `SCHEDULE` statement in the departure event places event notices on the future event list. If one understands the events that are necessary to simulate the M/M/1 queue, the rest of the model should be self-explanatory.

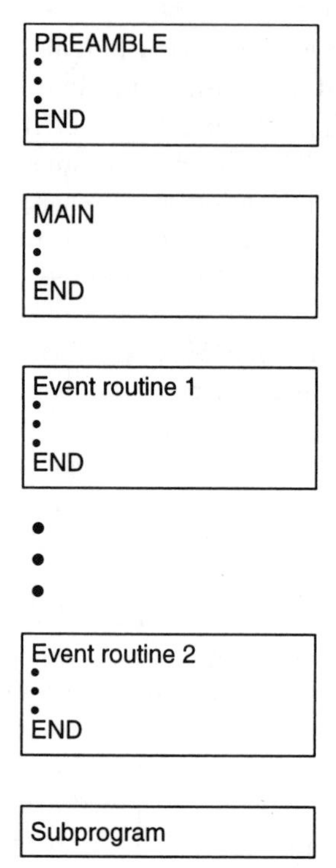

Figure 8.3 Program structure in SIMSCRIPT.

The major advantage of SIMSCRIPT is that its English-like structure aids in debugging and it provides some flexibility in model development. Besides this, it has several aids to debugging. Ironically, the most serious drawback lies with its structure. The syntax of SIMSCRIPT is uncontrolled; the language has essentially no formal structure. The use of optional words and keywords is inconsistent.

8.4 GASP

GASP is a FORTRAN-based simulation programming language developed by Alan Prisker in the early 1970s. GASP IV, the most popular version of GASP, is essentially a collection of about 34 FORTRAN subroutines designed for discrete, continuous, and hybrid simulations. In that sense, it is more of a simulation package than a simulation language. It is no use describing all the subroutines here, because a detailed description with numerous examples is provided by Prisker [13]. Basically, the GASP IV package is comprised of a time-advance subroutine called `GASP`, subroutines for managing a future event list, subroutines for data collection and statistical analysis, random variate generators, a report generator, and a subroutine, ERROR, to assist in debugging. The user must provide a main program with a statement CALL GASP to start the simulation, event routines, and a subroutine called EVNTS to handle all events.

Table 8.1
Typical declarations and statements in SIMSCRIPT [11].

Declaration/ Statement	Function	Remarks
PROCESS ...	Marks the beginning of process entity	Preamble declaration
TEMPORARY	Marks the beginning of resource entity	Preamble declaration
ENTITIES EVERY ...	Entity-attribute-set structure declarations	Preamble declaration
DEFINE ...	Defines properties of variables, routines, sets, etc.	Preamble declaration
MAIN	Marks the beginning of main segment in a program	Main declaration
PROCESS ...	Process routine heading	Routine declaration
LET ...	Assignment statement	Executable statement
FOR ...	Used for constructing program loops	Executable statement
START	Passes control to the timing mechanism	Executable statement
SIMULATION READ ...	Read data	I/O statement
PRINT ...	Print output with titles	I/O statement
CREATE ...	Create temporary entity	Entity management
FILE ...	Insert temporary entity into a set	Entity management
RELINQIUISH ...	Make resource available for reallocation	Process interactions
WAIT/WORK ...	Delay process execution	Process interactions
SCHEDULE ...	Create a new process	Process interactions

```
1  PREAMBLE
2     EVENT NOTICES INCLUDE ARRIVAL.GENERATOR AND DEPARTURE
3     DEFINE REPORT.GENERATOR AS ROUTINE
4     DEFINE  MEAN.INTERARRIVAL.TIME  AND  MEAN.SERVICE.TIME,  AND
      DELAY.IN.QUEUE AS REAL VARIABLES
5     DEFINE NUMBER.OF.CUSTOMERS AND NUMBER.OF.DEPARTURES AS
      INTEGER VARIABLES
6     DEFINE IDLE TO MEAN 0
7     DEFINE BUSY TO MEAN 1
8     TALLY AVERAGE.DELAY.IN.QUEUE AS THE AVERAGE AND
      NUMBER.OF.DELAYS AS THE NUMBER OF DELAY.IN.QUEUE
9  END

1  MAIN
2     LET MEAN.INTERARRIVAL.TIME = 1.0
3     LET MEAN.SERVICE.TIME = 0.5
4     LET NUMBER.OF.CUSTOMERS = 1000
5     LET SERVER = IDLE  ''  SO THAT THE FIRST ARRIVAL WILL FIND THE
                              SERVER IDLE
6     LET NUMBER.OF.DEPARTURES = 0
7     ACTIVATE AN ARRIVAL.GENERATOR
8     START SIMULATION
9  END

1  EVENT ARRIVAL.GENERATOR
2     WHILE TIME.V >= 0.0
3     DO
           WAIT EXPONENTIAL.F(MEAN.INTERARRIVAL.TIME,1) MINUTES
           ACTIVATE A CUSTOMER NOW
4     LOOP
5  RETURN
6  END

1  EVENT DEPARTURE
2     DEFINE TIME.OF.ARRIVAL AS A REAL VARIABLE
3     LET TIME.OF.ARRIVAL = TIME.V
4     LET DELAY.IN.QUEUE = TIME.V - TIME.OF.ARRRIVAL
5     ADD 1 TO NUMBER.OF.DEPARTURES
5     IF NUMBER.OF.DEPARTURES = NUMBER.OF.CUSTOMERS
           ACTIVATE A REPORT.GENERATOR
7     ALWAYS
8     IF QUEUE IS EMPTY
9          LET SERVER=IDLE
10    OTHERWISE
11    REMOVE FIRST CUSTOMER FROM QUEUE
12    WORK EXPONENTIAL.F(MEAN.SERVICE.TIME,2) MINUTES
13    SCHEDULE A DEPARTURE
14    REGARDLESS
15 RETURN
16 END

1  EVENT REPORT.GENERATOR
2     PRINT I LINE THUS
3          SIMULATION OF THE M/M/1 QUEUE
4     SKIP 2 OUTPUT LINES
5     PRINT 3 LINES WITH MEAN.INTERARRIVAL.TIME, MEAN.SERVICE.TIME,
6     AND NUMBER.OF.CUSTOMERS THUS
           MEAN INTERARRIVAL TIME             **.**
           MEAN SERVICE TIME                  **.**
           NUMBER OF CUSTOMERS SERVED         ****
7     SKIP 2 OUTPUT LINES
8     PRINT 2 LINES WITH AVERAGE.DELAY.IN.QUEUE AND
9     NUMBER.OF.DEPARTURES THUS
           AVERAGE DELAY IN QUEUE             **.**
           NUMBER OF DEPARTURES               ****
10 STOP
11 END
```

Figure 8.4a SIMSCRIPT program listing for the simulation of the M/M/1 queue.

```
SIMULATION OF THE M/M/1 QUEUE

MEAN INTERARRIVAL TIME              1.00
MEAN SERVICE TIME                    .50
NUMBER OF CUSTOMERS SERVED          1000


AVERAGE DELAY IN QUEUE               .43
NUMBER OF DEPARTURES                1000
```

Figure 8.4b Output of SIMSCRIPT model of the M/M/1 queue.

GASP IV can be used at any facility equiped with FORTRAN compiler. Analysts already proficient in FORTRAN can readily learn and use the language. GASP IV enjoys many of the advantages of its parent language, FORTRAN, but saves the analyst the time required for coding many of the common routines required for simulation. However, GASP IV suffers from the same deficiences as FORTRAN. The lack of an efficient list-processing capability makes queue handling inefficient. Some of these deficiencies have been overcome in several recent developments of GASP. These include GASP-PL/I, the PL/I version of GASP IV; GASP IV/E, an interactive version of GASP IV; and GASP V, which expands the continuous capabilities of GASP IV [14, 15].

8.5 SIMULA

SIMULA is an ALGOL-based simulation language designed and developed by Dahl and Nygaard [16] at the Norwegian Computing Center (NCC), Oslo, in early 1960s. As mentioned at the beginning of this chapter, any general-purpose language, including ALGOL, can be used by an analyst in simulation. SIMULA adds concepts particularly suitable for simulation work to ALGOL and, in that sense, can be regarded as a general-purpose language, a superset of ALGOL. SIMULA 67, the current version [17], contains ALGOL 60 as a subset.

The central idea of SIMULA is the concept of *object*, which is equivalent to an entity. An object reflects the features of the component, which are considered to be relevant to the purpose of the model. A class is a set of objects with similar characteristics. A good introduction is provided by Birtwistle *et al.* [18].

Although SIMULA is a relatively old and widely used language particularly in Europe and Australia, it has never gained any widespread use in commercial environments because of its complexity. Even to write a simple simulation program requires a working knowledge of ALGOL, which is not currently popular. SIMULA is not as good as it might be.

It is based on excellent ideas, some of which are poorly implemented. However, SIMULA has had a great impact on the design of programming languages. Languages such as CONCURRENT PASCAL, ACT, and SMALLTALK are rooted in concepts from SIMULA.

8.6 SLAM

SLAM, Simulation Language for Alternative Modeling, is a FORTRAN-based simulation language designed by Dennis Pegden in 1979 and subsequently supported and marketed by Pritsker and Associates, Inc. SLAM was based on GASP IV and Q-GERT (an acronym for Queues—Graphical Evaluation and Review Techniques) and has three structures: process, event, and continuous. The process structure is similar to Q-GERT, and the event structure is similar to GASP IV. Like a GPSS model, a SLAM model requires fewer statements than a FORTRAN model.

SLAM II, the current version of SLAM, is available in different forms. The basic SLAM II does not include animation. SLAMSYSTEM, the microcomputer version of SLAM II, has animation, graphics capabilities, and a user-friendly environment. SLAM II/TESS is similar to SLAMSYSTEM but, in addition, has enhanced statistical features and an integrated database for model input/output. MHEX (Material Handling Extension) is an extension to SLAM II with facilities for handling materials and simulating vehicle and retrieval systems.

Using the process structure of SLAM requires that the analyst first develops a graphical network representing the flow of an entity (or object) in the system. The network model represents all possible paths that the entity can take as it flows through the system. The network is developed by combining a standard set of *nodes* and *branches*. A node may represent an event or system action such as an arrival event and queue, whereas a branch is used to represent an activity or the passage of time, such as service and time between creation of entities. The analyst directly translates the network representation of the system into computer statements. Alternatively, the analyst could develop the computer statements directly without using a network model of the system. The analyst codes each discrete event as a FORTRAN subroutine. FORTRAN has subprograms for performing common simulation features such as random variate generation, event scheduling, and statistics collection. The subroutine SLAM is called by the user-written main program and serves as the executive control for a discrete event simulation. It automatically handles event-handling algorithms, random variate generation, and routines for gathering statistics.

A good introduction to SLAM is given by Pritsker [19]. Pritsker and Associates give short courses on the use of SLAM.

8.7 RESQ

RESQ (RESearch Queueing) package, a product of IBM research, is a software tool for building queueing network models. A main feature of the simulation language is the ability to describe models in a hierarchical manner, thereby permitting the user to define submodels to be applied in a fashion similar to the use of macros in a programming language.

Like other simulation languages, the development of RESQ has been evolutionary. Version 2 of RESQ is a system for constructing and solving queueing networks. The class of networks that can be simulated by RESQ includes general multiserver queues, passive queues, and complex routine decisions. Queueing networks can be defined, listed, evaluated, and reconstructed either interactively or by writing programs that call RESQ routines. RESQ allows explicit consideration of many system features that are usually ignored in queueing models.

The constructs of RESQ are oriented toward computer and communication system features. The elements of RESQ include [20]:

1. A population of jobs. Each job has an attached variable, which can be used to retain job attributes.
2. A set of queues, which may be *active* or *passive*. An active queue is a queue in the traditional sense. It consists of a set of servers, a set of waiting rooms for customers waiting for or receiving service, and a control mechanism for service discipline. A passive queue consists of a pool of tokens, a nonempty set of waiting areas for customers possessing or requesting tokens, and a control mechanism for the tokens and customers.
3. A set of nodes, which may be parts of queues or used for auxiliary functions.
4. A set of routing rules, which allow probabilistic and deterministic routing of customers or packages in the network.

Effective use of RESQ is based on constructing diagrams representing queueing network models. A typical queueing network model of an interactive computer system is presented in Figure 8.5. Additional technical description of RESQ and queueing network modeling may be found in Sauer *et al.* [21]. Introductory material on RESQ with examples can be found in Sauer [22], Sauer and MacNair [23], and MacNair and Sauer [24]. RESQME (RESeach Queueing Modeling Environment), a recent version of RESQ, is a workstation environment [25]. It is an integrated, graphics-oriented system of hardware and software, designed to facilitate the use of models for performance evaluation purposes.

8.8 Criteria for Language Selection

There are two types of factors that influence the selection of the special-purpose language an analyst uses in his simulation. One set of factors is concerned with the operational characteristics of the language; the other is related to its problem-oriented characteristics [26, 27].

In view of the operational characteristics of a language, an analyst must consider factors such as the following:

1. The analyst's familiarity with the language
2. The ease with which the language can be learned and used if the analyst is not familiar with it
3. The languages supported at the installation where the simulation is to be done
4. The complexity of the model
5. The need for a comprehensive analysis and display of simulation results
6. The language's provision of good error diagnostics
7. The compiling and running time efficiency of the language
8. The availability of a well-written user's manual
9. The availability of support by a major interest group
10. The cost of installation, maintenance, and updating of the language

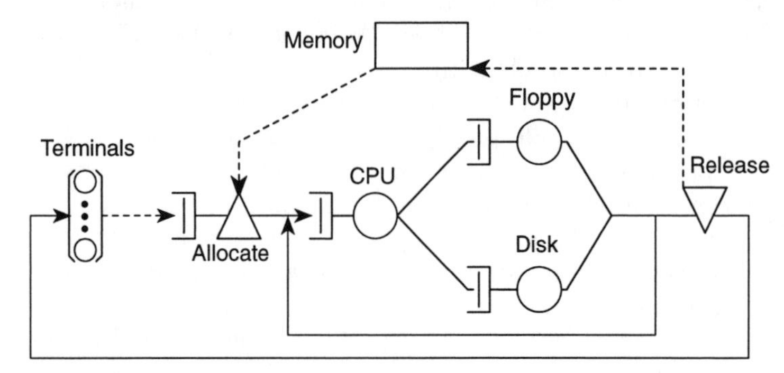

Figure 8.5 A typical RESQ queueing network model.

In viewing the characteristics of the language and the problems the analyst will most likely encounter, the following factors should be considered:

1. Time advance methods
2. Random number and random variate generation capabilities.
3. The way a language permits the analyst to control the sequence of subroutines that represent the state changes
4. Capability for inserting user-written subroutines
5. Forms of statistical analyses that can be performed on the data collected during simulation

No language is without some strong points as well as weak points. It is difficult to compare these languages because many important software features are quite subjective in nature. Everyone seems to have his own opinion concerning the desirable features of a simulation language—e.g., ease of model development, availability of technical assistance, and system compatibility. In spite of this difficulty, various attempts have been to compare simulation languages based on objective criteria [4, 12].

In general, GPSS and SLAM (which are FORTRAN based) are easiest to learn. SIMSCRIPT has the most general process approach and thus can be used to model any system without using the event-scheduling approach. However, this may result in more lines of code than GPSS or SLAM. SIMULA has the most adequate formal syntactic structure, for efficient compilation and programming ease. RESQ has features specially oriented toward computer and communication systems, but the terminology is strictly in terms of queueing networks.

8.9 Summary

This chapter has provided a brief introduction to six simulation languages: GPSS, SIMSCRIPT, GASP, SIMULA, SLAM, and RESQ. GPSS, SIMSCRIPT, GASP, and SIMULA have been in use longer than SLAM and RESQ. Criteria for deciding which of these languages an analyst selects in developing his model have been discussed.

In addition to these languages, there are several others of note. SIMAN was written by Dennis Pegden [28] (who also wrote SLAM) and introduced in 1983. It is a combined discrete continuous simulation language based on FORTRAN for modeling general systems with special features for manufacturing systems. SIMNET, a network-based simulation language coded in FORTRAN 77, was developed by Hamdy Taha [29] and introduced in 1988. Although the language does not allow the use of external FORTRAN inserts (subroutines), it is capable of modeling extremely complex situations. SMPL is an older but less popular simulation language especially useful for communication networks [30]. NETWORK II.5, COMNET II.5, and LANNET II.5 are three new simulators for performance evaluation of computer systems and LANs [31]. No programming is required in any of the three. All the analyst has to do is graphically define network topology through routing nodes and transmission links; animation follows immediately. NETWORK II.5 and LANNET II.5 are mainly for LANs; COMNET can handle wide area networks.

The development of new simulation languages has slowed considerably in the last few years, and the well established languages have not changed dramatically for the past few years. This notwithstanding, it is expected that new languages will be developed and old ones will be improved. At present there is a growing interest in combined discrete continous simulations. Also, the use of ADA and C as simulation languages is receiving active attention [11, 32].

In Table 8.2, we present the addresses of vendors that distribute common simulation languages. The vendors of the major simulation languages all claim considerable user bases and offer some support to users.

8.10 References

[1] G. S. Fishman, *Concepts and Methods in Discrete Event Digital Simulation.* New York: John Wiley & Sons, 1973, pp. 92–96.

[2] G. Gordon, "A General Purpose Systems Simulation Program", *Proc. EJCC*, Washington DC: Macmillan, 1961.

[3] J. Banks, J. S. Carson, and J. N. Sy, *Getting Started with GPSS/H.* Annandale, VA: Wolverine Software, 1989.

[4] Minuteman Software, *GPSS/PC Manual.* Stow: MA, 1988.

[5] B. Schmidt, *The Simulator GPSS-FORTRAN Version 3.* New York: Springer Verlag, 1987.

[6] T. J. Schriber, *Simulation Using GPSS.* New York: John Wiley & Sons, 1974.

[7] K. A. Dunning, *Getting Started in GPSS.* San Jose, CA: Engineering Press, 1985.

[8] P. A. Bobillier, B. C. Kahan, and A. R. Probst, *Simulation with GPSS and GPSS V.* Englewood Cliffs, NJ: Prentice-Hall, 1976.

[9] H. M. Markowitz, B. Hausner, and H. W. Karr, *SIMSCRIPT: A Simulation Programming Language.* Englewood Cliffs, NJ: Prentice-Hall, 1963.

[10] P. J. Kiviat, R. Villaneuva, and H. M. Markowitz, *The SIMSCRIPT II Programming Language.* Englewood Cliffs, NJ: Prentice-Hall, 1968.

[11] F. Neelamkavil, *Computer Simulation and Modelling.* Chichester, U.K.: John Wiley & Sons, 1987, pp. 210, 292–294.

[12] A. M. Law and W. D. Kelton, *Simulation Modeling and Analysis.* New York: McGraw-Hill, 2nd ed., 1991, pp. 254–258.

[13] A. A. B. Pritsker, *The GASP IV Simulation Language.* New York: John Wiley & Sons, 1974.

[14] A. A. B. Pritsker and R. E. Young, *Simulation with GASP-PL/I.* New York: John Wiley & Sons, 1975.

[15] F. Cellier and A. E. Blitz, "GASP V: A Universal Simulation Package," *Proc. IFAC Conference*, 1976.

[16] O. J. Dahl and K. Nygaard, "SIMULA—An ALGOL-based Simulation Language," *Comm. ACM*, vol. 9, no. 9, Sept. 1966, pp. 671–678.

[17] O. J. Dahl, B. Myhrhang, and K. Nygaard, *Common Base Language*, Norwegian Computing Center, Oslo, October 1970.

[18] G. M. Birtwistle *et al.*, *SIMULA BEGIN.* Philadelphia: Auerbach Publishers, 1973.

[19] A. A. B. Pritsker, *Introduction to Simulation and SLAM II.* New York: John Wiley & Sons, 2nd ed., 1984.

[20] C. H. Sauer *et al.*, "RESQ—A Package for Solution of Generalized Queueing Networks," *Proc. 1977 National Computer Conf.*, 1977, pp. 977–986.
[21] ———, "The Research Queueing Package: Past, Present and Future," *Proc. 1982 National Computer Conf.*, 1982.
[22] ———, "The Research Queueing Package Version 2: Introduction and Examples," *IBM Research Report RA-138*, Yorktown Heights, NY, April 1982.
[23] C. H. Sauer and E. A. MacNair, *Simulation of Computer Communication System.* Englewood Cliffs, NJ: Prentice-Hall, 1983.
[24] E. A. MacNair and C. H. Sauer, *Elements of Practical Performance Modeling.* Englewood Cliffs, NJ: Prentice-Hall, 1985.
[25] J. F. Kurose *et al.*, "A graphics-oriented modeler's workstation environment for the RESearch Queueing Package (RESQ)," *Proc. ACM/IEEE Fall Joint Computer Conf.*, 1986.
[26] W. J. Graybeal and U. W. Pooch, *Simulation: Principles and Methods.* Cambridge, MA: Winthrop Publishers, 1980, p. 153.
[27] J. R. Emshoff and R. L. Sisson, *Design and Use of Computer Simulation Models.* London: Macmillan, 1970, pp. 119–150.
[28] C. D. Pegden, *Introduction to SIMAN with Version 3.0 Enhancements.* State College, PA: Systems Modeling Corporation, 1985.
[29] H. A. Taha, *Simulation Modeling and SIMNET.* Englewood Cliffs, NJ: Prentice Hall, 1988.
[30] M. H. MacDougall, *Simulating Computer Systems: Techniques and Tools.* Cambridge, MA: MIT Press, 1987.
[31] CACI Products Company, "Six Simulation Solutions," *IEEE Spectrum*, May 1991, p. 1.
[32] M. Pidd (ed.), *Computer Modelling for Discrete Simulation.* Chichester, UK: John Wiley & Sons, 1989, pp. 217–240.

Problems

8.1 What is a simulation programming language?
8.2 Compare and contrast general-purpose languages and special-purpose languages.
8.3 Find out from the GPSS manual how the clock is updated for the next event time.
8.4 Compare and contrast GPSS and SIMSCRIPT, the two popular discrete system simulation languages.
8.5 Find out the programming languages available at your computer center. Which one would you prefer for writing simulation? Why?

Table 8.2
List of Addresses of Language Vendors.

Vendor Address	Language
CACI Products Company 3344 North Torrey Pines Court La Jolla, CA 92037 (619) 457-9681	SIMSCRIPT II.5, NETWORK II.5, COMNET II.5, LANNET II.5
Prisker Corporation P.O. Box 2413 1305 Cumberland Avenue West Lafayette, IN 47906 (800) 537-9221 or 8910 Purdue Road, Suite 500 Indianapolis, IN 46268 (317) 879-1011	GASP, SLAM II
Control Data Corp. 8100 34th Avenue S. Minneapolis, MN	GPSS III, SIMSCRIPT
IBM 112 E. Post Road White Plains, NY 10600	GPSS/360, SIMSCRIPT
SIMULA P.O. Box 4403 Torskov, N-0402 Oslo 4, Norway (412) 156-710	SIMULA
SimTec, Inc. P.O. Box 3492 Fayetteville, Arkansas 72702	SIMNET
Edward A. MacNair IBM Hawthorne Research Laboratory P.O. Box 704 Yorktown Heights, NY 10598 (914) 784–7561	RESQ II

Selected Bibliography

A. LOCAL AREA NETWORKS

Ahuja, V., *Design and Analysis of Computer Communications Networks.* New York: McGraw-Hill, 1982.

Brooks, T. (ed.), *The Local Area Network Reference Guide.* Englewood Cliffs: Prentice-Hall, 1985.

Claiborne, J. D., *Mathematical Preliminaries for Computer Networking Performance.* New York: John Wiley & Sons, 1990.

Cooper, R. B., *Introduction to Queuing Theory.* New York: North-Holland, 2nd ed., 1981.

Fortier, P. J. (ed.), *Handbook of LAN Technology.* New York: McGraw-Hill, Intertext Publication, 1992.

Fortier, P. J., and Desrochers, G.R., *Modeling and Analysis of Local Area Networks.* Boca Raton, FL: CRC Press, 1990.

Hammond J. L., and O'Reilly P. J. P., *Performance Analysis of Local Computer Networks.* Reading, MA: Addison-Wesley, 1986.

Hasegawa, T. *et al.* (eds.), *Computer Networking and Performance.* Amsterdam: North-Holland, 1986.

Hayes, J. F., *Modeling and Analysis of Computer Communications Networks.* New York: Plenum Press, 1984. Computer networks, mathematical models.

King, P. J. B., *Computer and Communication Systems Performance Modelling,* Herfordshire, UK: Prentice-Hall, 1990.

Kleinrock, L., *Queueing Systems.* New York: John Wiley & Sons, 1975.

Kobayashi, H., *Modeling and Analysis: An Introduction to System Performance Evaluation Methodology.* Reading, MA: Addison-Wesley, 1978.

Mitrani, I., *Modelling of Computer and Communication Systems.* Cambridge, UK: Cambridge University Press, 1987.

Nussbaumer, H. *Computer Communication Systems: Principles Design Protocols.* Vol. II New York: John Wiley & Sons, 1990.

Ravasio, P. C., *et al.* (eds.), *Local Computer Networks.* New York: North-Holland, 1982.

Sauer, C. H., and MacNair, E. A., *Simulation of Computer Communication Systems.* Englewood Cliffs, NJ: Prentice-Hall, 1983.

Schoemaker, S. (ed.), *Computer Networks and Simulation III.* Amsterdam: North- Holland 1986.

Schwartz, M., *Computer Communications Network Design and Analysis.* Englewood Cliffs, NJ: Prentice-Hall, 1977.

Schwartz, M., *Telecommunication Networks, Protocols, Modeling and Analysis.* Reading, MA: Addison-Wesley, 1987.

Slone, J. P., and Drinan, A. (eds.), *Handbook of Local Area Networks.* Boston: Auerbach Publishers, 1991.

Stallings, W., *Data and Computer Communications,* New York: Macmillian, 1985.

——, *Data and Computer Communications,* New York: Macmillian, 2nd ed., 1988.

——, *Local Networks and Metropolitan Area Networks.* New York: Macmillian, 4th ed., 1993.

Stuck B. W. and Arthurs, E., *A Computer and Communications Network Performance Analysis Primer.* Englewood Cliffs: Prentice-Hall, 1985.

Tagaki, H., *Analysis of Polling Systems.* Cambridge, MA: MIT Press, 1986.

Tanenbaum, A. S., *Computer Networks.* Englewood Cliffs, NJ: Prentice-Hall, 1989.

Walrand, J., *Communication Networks: A First Course.* Homewood, IL: Irwin & Aksen, 1991.

B. SIMULATION

Banks, J. and Carson J. S., II. *Discrete Event Systems Simulation.* Englewood Cliffs, NJ: Prentice-Hall, 1984.

Bobillier, P. A., Kahan, B. C., and Probst, A. R. *Simulation with GPSS and GPSS V.* Englewood Cliffs, NJ: Prentice-Hall, 1976.

Birtwistle, G. M. *et al.*, *SIMULA BEGIN.* Philadelphia: Auerbach/Studentliteratur, 1973.

Bratley, P., Fox, B. L. and Schrage, L. E. *A Guide to Simulation.* New York/Heidelberg: Springer-Verlag, 2nd ed, 1987.

Deo, N. *System Simulation with Digital Computers.* Englewood Cliffs, NJ: Prentice-Hall (Indian Edition), 1979.

Dunning, K. A., *Getting Started in GPSS.* San Jose, CA: Engineering Press, 1985.

Emshoff, J. R. and Sisson, R. L. *Design and Use of Computer Simulation Models.* London: Macmillan, 1970.

Fishman, G. S. *Concepts and Methods in Discrete Event Digital Simulation.* New York: John Wiley & Sons, 1973.

——, *Principles of Discrete Event Simulation.* New York: John Wiley & Sons, 1978.

Franta, W. R. *The Process View of Simulation.* New York/Amsterdam: Elsevier, 1977.

Gordon, G. *System Simulation.* Englewood Cliffs, NJ: Prentice-Hall, 2nd ed, 1978.

Graybeal, W. J. and Pooch, U. W. *Simulation: Principles and Methods.* Cambridge, MA: Winthrop, 1980.

Hollingdale, S. H. (ed.) *Digital Simulation in Operation Research.* New York: Elsevier, 1967.

Jain, R., *The Art of Computer Systems Performance Analysis.* New York: John Wiley & Sons, 1991.

Kiviat, P. J., Villaneuva, R., and Markowitz, H. M., *The SIMSCRIPT II Programming Language.* Englewood Cliffs, NJ: Prentice-Hall, 1968.

Law, A. and Kelton, W. D. *Simulation Modelling and Analysis.* New York: McGraw-Hill, 2nd ed., 1991.

Lehman, R. S. *Computer Simulation and Modelling: An Introduction.* Hillsdale, NJ: Lawrence Erlbaum, 1977.

Lewis, T. G. and Smith, B. J. *Computer Principles of Modelling and Simulation.* Boston: Houghton and Mifflin, 1979.

MacDougall, M. H., *Simulating Computer Systems: Techniques and Tools.* Cambridge, MA: MIT Press, 1987.

MacNair, E. A., and Saucer, C. H. *Elements of Practical Performance Modelling.* Englewood Cliffs, NJ: Prentice-Hall, 1985.

Maisel, H. and Gnugnoli, C., *SIMULATION of Discrete Stochastic Systems.* Chicago: Science Research Associates, 1972.

Markowitz, H. M., Hausner, B., and Karr, H. W. *SIMSCRIPT: A Simulation Programming Language.* Englewood Cliffs, NJ: Prentice-Hall, 1963.

Martin, F. R., *Computer Modelling and Simulation.* New York: John Wiley & Sons, 1968.

Maryanski, F., *Digital Computer Simulation.* Rochelle Park, NJ: Hayden, 1980.

Meier, R. C., Newell, W. T., and Pazer, H. L., *Simulation in Business and Economics.* Englewood Cliffs, NJ: Prentice-Hall, 1969.

Mihram, A. G., *Simulation: Statistical Foundations and Methodology.* New York: Academic Press, 1972.

Mitrani, I., *Simulation Techniques for Discrete Event Systems.* Cambridge, UK: Cambridge University Press, 1982.

Naylor, T. H. (ed.) *Computer Simulation with Models of Economic Systems.* New York: John Wiley & Sons, 1971.

Naylor, T. H., Balintfly, J. L., Burdick, D. S., and Chu, K., *Computer Simulation Techniques.* New York: John Wiley & Sons, 1966.

Neelamkavil, F. *Computer Simulation and Modelling.* Chichester, UK: John Wiley & Sons, 1987.

Payne, J.A., *Introduction to Simulation.* New York: McGraw-Hill, 1982.

Pegden, C. D., *Introduction to SIMAN with Version 3.0 Enhancements.* State College, PA: Systems Modeling Corporation, 1985.

Pidd M. (ed.), *Computer Modelling for Discrete Simulation.* Chichester, UK: John Wiley & Sons, 1989.

Pritsker, A. A. B., *The GASP IV Simulation Language.* New York: John Wiley & Sons, 1974.

——, *Simulation and SLAM II.* New York: John Wiley & Sons, 2nd eds., 1984.

——, *Introduction to Simulation and SLAM II.* New York: John Wiley & Sons, 2nd ed., 1984.

Reitman, J., *Computer Simulation Applications.* New York: John Wiley & Sons, 1971.

Rubinstein, R. J. *Simulation and the Monte Carlo Method.* New York: John Wiley & Sons, 1981.

Sauer, C. H., and Chandy K. M. *Computer Systems Performance Modelling.* Englewood Cliffs, NJ: Prentice-Hall, 1981.

Schmidt, J. W. and Taylor, R. E. *Simulation and Analysis of Industrial Systems.* Homewood: Irwin, 1970.

Schmidt, B., *The Simulator GPSS-FORTRAN Version 3.* London/New York: Springer-Verlag, 1987.

Schriber, T. J., *Simulation Using GPSS.* New York: John Wiley & Sons, 1974.

Shannon, R. E., *System Simulation—the Art and Science.* Englewood Cliffs, NJ: Prentice-Hall, 1975.

Solomon, S. L., *Simulation of Waiting Line Systems.* Englewood Cliffs, NJ: Prentice-Hall, 1983.

Taha, H. A., *Simulation Modelling and SIMNET.* Englewood Cliffs, NJ: Prentice-Hall, 1988.

Tocher, K. D., *The Art of Simulation.* New York: Van Nostrand, 1963.

Watson, H. J., *Computer Simulation in Business.* New York: John Wiley & Sons, 1981.

Zeigler, B. P., Elzas, M. S., Klir, G. J. and Oren, T. I. (eds.), *Methodology in Systems Modelling and Simulation.* Amsterdam/New York: North-Holland, 1979.

Zeigler, B. P., *Theory of Modelling and Simulation.* New York: John Wiley & Sons, 1976.

——, *Multifacetted Modelling and Discrete Event Simulation.* London/Orlando: Academic Press, 1984.

Appendix A

Simulation for Token-Passing LANs

```
/* This program simulates token-passing ring and bus LANs.  */

# include     <stdio.h>
# include     <stdlib.h>
# include     <math.h>

# define      MAX_STATIONS      50             /* Maximum number of stations */
# define      RING_OR_BUS       1              /* Flag to choose
                                               ring or bus LAN */
# define      RATE              10000000.0     /* Transmission rate in bps */
# define      PACKET_LENGTH     1000.0         /* Packet length in bits */
# define      MEDIUM_LENGTH     2000.0         /* Medium length in meters */
# define      MAX_PACKETS       10000          /* Maximum packets to be
                                               transmitted in a simulation
                                               run */
# define      FACTOR            1000.0         /* A factor used for changing
                                               units of time */
# define      TOKEN_LENGTH      10.0           /* Token length in bits */
# define      STN_LATENCY       1.0            /* Station latency in bits */
# define      MAX_Q_SIZE        100            /* Maximum queue size */
# define      DEGREES_FR        5              /* Degrees of freedom */

float arrival_rate; /* Arrival rate (in packets/sec) per station */
float packet_time; /* Packet transmission time */
float stn_latency; /* Station latency in time units */
float token_time; /* Token transmission time */
float tau; /* End-to-end propagation delay */
float t_dist_par[10] = {12.706, 4.303, 3.182, 2.776, 2.571, 2.447, 2.365,
2.306, 2.262, 2.228}; /* T-distribution parameters */
float start_time[MAX_STATIONS][MAX_Q_SIZE]; /* Starting time of packet */
float event_time[MAX_STATIONS][3]; /* Time of occurrence of an event */
float delay_ci[DEGREES_FR + 1]; /* Array to store delay values */
```

```
float rho, clock, no_pkts_departed, next_event_time;
float x, logx, flag, infinite, rand_size;
float delay, total_delay, average_delay, walk_time;
float delay_sum, delay_sqr, delay_var, delay_sdv, delay_con_int;

int queue_size[MAX_STATIONS]; /* Current queue size at a station */
int next_stn[MAX_STATIONS]; /* Array to identify the next station */
int previous_stn[MAX_STATIONS]; /* Array to identify previous station */
int in[MAX_STATIONS]; /* Array to identify status of a station */

int i, j, ic, ii, temp_flag, next, next_station, next_event;
int ring_size, ring_or_bus, stn_to_add, temp_stn;

main ()
{
printf("The following results are for:  \n");
printf("Degrees of freedom = % d\n", DEGREES_FR);
printf("Confidence interval = 95 percent \n");
printf(" ========================================== \n");
printf("\n");
arrival_rate     = 0.0;
packet_time      = PACKET_LENGTH * FACTOR / RATE;
stn_latency      = STN_LATENCY * FACTOR / RATE;
token_time       = TOKEN_LENGTH * FACTOR / RATE;
tau              = MEDIUM_LENGTH * FACTOR * 5.0 * pow (10.0, -9.0);
infinite         = 1.0 * pow (10.0,30.0);
ring_or_bus      = RING_OR_BUS;
rand_size        = 0.5 * pow (2.0, 8.0 * sizeof(int));

for (ii = 0; ii < 10; ii++)
        {
        arrival_rate = arrival_rate + 20.0;
        for (ic = 0; ic <= DEGREES_FR; ic++)
         {
         rho                  = 0.0;
         clock                = 0.0;
         no_pkts_departed     = 0.0;
         total_delay          = 0.0;
         next_event_time      = 0.0;
         average_delay        = 0.0;
         flag                 = 1.0;
        /* Compute the traffic intensity.  If the traffic intensity
        is greater than unity, stop the program.  */

         rho = arrival_rate * PACKET_LENGTH * MAX_STATIONS / RATE;

         if (rho >= 1.0)
           {
           printf("Traffic intensity is too high");
           exit(1);
           }
         /* Initialize all variables to their appropriate values.  */

           for (i = 0; i < MAX_STATIONS; i++)
           {
           queue_size[i] = 0;
           for (j = 0; j < MAX_Q_SIZE; j++)
             {
             start_time[i][j] = 0.0;
             }
           }
```

```
if (ring_or_bus == 1)       /* This is for ring LANs.  */
  {
  ring_size = MAX_STATIONS;
  walk_time = token_time + stn_latency + tau/MAX_STATIONS;
  }
else                        /* This is for bus LANs.  */
  {
  ring_size = 0;
  walk_time = token_time + tau/3.0;
  }

for (i = 0; i < MAX_STATIONS; i++)
  {
  if (ring_or_bus == 1)   /* For ring LANs */
    {
    next_stn [MAX_STATIONS-1] = 0;
    previous_stn [0] = MAX_STATIONS-1;
    previous_stn [MAX_STATIONS-1] = MAX_STATIONS-2;
    next_stn [0] = 1;
    if ((i < (MAX_STATIONS-1) && (i > 0)))
     {
     next_stn [i] = i+1;
     previous_stn [i] = i-1;
     }
    }
  else                  /* For bus LANs */
    {
    in[i] = 0;
    next_stn [i] = -1;
    previous_stn [i] = -1;
    }
  }

for (i = 0; i < MAX_STATIONS; i++)
  {
  for (j = 0; j < 3; j++)
    {
    event_time[i][j] = 0.0;
    if (j != 0) event_time[i][j] = infinite;
    }
  }
/* Scan the event list and pick the next event to be executed.  */

while (no_pkts_departed < MAX_PACKETS)
  {
  next_event_time = infinite;
  for (i = 0; i < MAX_STATIONS; i++)
    {
    for (j = 0; j < 3; j++)
     {
     if (next_event_time > event_time[i][j])
      {
      next_event_time = event_time[i][j];
      next_station = i;
      next_event = j;
      }
     }
    }
  clock = next_event_time;            if (next_event > 2)
    {
    printf("Check the event-list");
    exit(1);
    }
  switch (next_event)
```

```
{
case 0:    /* This is an arrival event.  */
 {
 queue_size[next_station] ++ ;
 if (queue_size[next_station] > MAX_Q_SIZE)
  {
  printf("The queue size is large and is = %d\n",
  queue_size[next_station]);
  exit(1);
  }
 if (ring_or_bus == 1) /* This is for a token-passing ring.  */
  {
  if (flag == 1.0)
   {
   flag = 0.0;
   event_time[next_station][2] = clock;
   }
  }
else   /* This is for a token-passing bus.  */
   {
  if (flag == 1.0)
   {
   flag = 0.0;
   ring_size = 1;
   in[next_station] = 1;
   next_stn [next_station] = next_station;
   previous_stn [next_station] = next_station;
   event_time[next_station][2] = clock;
   }
  }

/* Schedule the next arrival.  */
for (;;)
  {
  x = (float) rand();
  if (x != 0.0) break;
  }
logx = -log(x/rand_size) * FACTOR / arrival_rate;
event_time[next_station][next_event] = clock + logx;
start_time [next_station][queue_size[next_station]-1] = clock;
break;
}

case 1:    /* This is a departure event.  */
 {
 queue_size[next_station] -- ;
 no_pkts_departed ++ ;
 delay = clock - start_time [next_station][0];
 total_delay += delay;

 /* Push the queue forward.  */

 for (i = 0; i < queue_size[next_station]; i++)
   start_time[next_station][i] = start_time[next_station][i+1];
 start_time[next_station][queue_size[next_station]] = 0.0;
 event_time[next_station][next_event] = infinite;
 if (ring_or_bus == 0) /* For bus LANs */
   {
   stn_to_add = -1;
   for (i = next_station + 1; i < MAX_STATIONS; i++)
    {
    if ((queue_size[i] > 0) && (in[i] == 0)) stn_to_add = i;
    if (stn_to_add != -1) continue;
    }
```

```
if (stn_to_add == -1)
 {
 for (i = 0; i < next_station - 1; i++)
  {
  if ((queue_size [i] > 0) && (in[i] == 0)) stn_to_add = i;
  if (stn_to_add != -1) continue;
  }
 }
  if (stn_to_add != -1)
   {
   temp_stn = next_stn[next_station];
   next_stn[next_station] = stn_to_add;
   next_stn[stn_to_add] = temp_stn;
   previous_stn[stn_to_add] = next_station;
   previous_stn[temp_stn] = stn_to_add;
   ring_size ++ ;
   in[stn_to_add] = 1;
   }
  if (queue_size[next_station] == 0)
   {
   ring_size -- ;
   in[next_station] = 0;
   if (ring_size == 0)
     {
     next_stn[next_station] = -1;
     previous_stn[next_station] = -1;
     flag = 1.0;
     }
   else
     {
     next = next_stn[next_station];
     event_time[next][2] = clock + walk_time;
     next_stn[previous_stn[next_station]] =
      next_stn[next_station];
     previous_stn[next] = previous_stn[next_station];
     }
   }
 else
   {
   next = next_stn[next_station];
   event_time[next][2] = clock + walk_time;
   }
 }
if (ring_or_bus == 1)   /* For ring LANs */
 {
 next = next_stn[next_station];
 if ((next == 0) && (queue_size[next_station] == 0))
{
temp_flag = 1;
for (i = 0; i < MAX_STATIONS; i++)
     {
     if (queue_size[i] != 0)
      {
      event_time[next][2] = clock + walk_time;
      temp_flag = 0;
      break;
      }
     }
```

```
          if (temp_flag == 1)
            {
            flag = 1.0;
            event_time[next][2] = infinite;
            }
          }
        else
          {
          event_time[next][2] = clock + walk_time;
          }

      break;
      }
      case 2:     /* This is a token arrival event.  */
       {
       event_time[next_station][2] = infinite;
       if (queue_size[next_station] > 0)
        {
        event_time[next_station][1] = clock + packet_time;
        }
       else
        {
        if (ring_or_bus == 0) /* For bus LANs */
          {
          printf("There is something wrong (bus LAN)");
          }
        else /* For ring LANs */
          {
          next = next_stn[next_station];
          if ((next == 0) && (queue_size[next] == 0))
            {
            temp_flag = 1;
            for (i = 0; i < MAX_STATIONS; i++)
             {
             if (queue_size[i] != 0)
              {
              event_time[next][2] = clock + walk_time;
              temp_flag = 0;
              break;
              }
             }
          if (temp_flag == 1)
            {
            flag = 1.0;
            event_time[next][2] = infinite;
            }
          }
        else
          {
          event_time[next][2] = clock + walk_time;
          }
         }
        }
      break;
      }
    }
  }
average_delay = total_delay / (no_pkts_departed * FACTOR);
delay_ci[ic] = average_delay
}
```

```
        delay_sum = 0.0;
        delay_sqr = 0.0;
        for (ic = 0; ic <= DEGREES_FR; ic++)
         {
         delay_sum += delay_ci[ic];
         delay_sqr += pow (delay_ci[ic],2.0);
         }
        delay_sum = delay_sum / (DEGREES_FR + 1);
        delay_sqr = delay_sqr / (DEGREES_FR + 1);
        delay_var = delay_sqr - pow(delay_sum,2.0);
        delay_sdv = sqrt(delay_var);
        delay_con_int = delay_sdv * t_dist_par[DEGREES_FR-1]/sqrt (DEGREES_FR);
        printf("For an arrival rate = %g\n",arrival_rate);
        printf("The average delay = %g", delay_sum);
        printf(" +- %g\n", delay_con_int);
        printf("\n");
    }
  }
```

Appendix B

Simulation Program for CSMA/CD LANs

```
/* This program simulates CSMA/CD local area networks.  */

# include     <stdio.h>
# include     <stdlib.h>
# include     <math.h>


# define    MAX_STATIONS      10            /* Number of stations */
# define    BUS_RATE          2000000.0     /* Transmission rate in bps */
# define    PACKET_LENGTH     1000.0        /* Packet length in bits */
# define    BUS_LENGTH        2000.0        /* Bus length in meters */
# define    MAX_BACKOFF       15.0          /* Backoff period in slots */
# define    PERSIST           0.5           /* Persistence */
# define    JAM_PERIOD        5.0           /* Jamming period */
# define    MAX_PACKETS       10000         /* Maximum packets to be trans-
                                            mitted in a simulation run */
# define    FACTOR            1000.0        /* A factor used for changing
                                            units of time */
# define    MAX_Q_SIZE        500           /* Maximum queue size */
# define    ID_SIZE           5000          /* Size of the identity array */
# define    DEGREES_FR        5             /* Degrees of freedom */

float arrival_rate; /* Arrival rate (in packets/sec) per station */
float arrival_rate_slots; /* Arrival rate (in packets/slot) per station */
float packet_time; /* Packet transmission time */
float t_dist_par[10] = {12.706, 4.303, 3.182, 2.776, 2.571, 2.447, 2.365,
                 2.306, 2.262, 2.228}; /* T-distribution parameters */
float start_time[ID_SIZE]; /* Starting time of packet */
float event_time[MAX_STATIONS][4]; /* Time of occurrence of an event */
float delay_ci[DEGREES_FR+1]; /* Array to store delay values */
float utilization_ci[DEGREES_FR+1]; /* Array for utilization values */
float throughput_ci[DEGREES_FR+1]; /* Array to store throughput values */
float collision_rate_ci[DEGREES_FR+1]; /* Array to save collision rates */


float slot_size, p, ch_busy;
float rho, clock, d_clock, no_pkts_departed, next_event_time;
float x, logx, rand_size, infinite;
float delay, total_delay, average_delay;
float delay_sum, delay_sqr, delay_var, delay_sdv, delay_con_int;
```

```
float utilization, utilization_sum, utilization_sqr;
float utilization_var, utilization_sdv, utilization_con_int;
float throughput, throughput_sum;
float collision_rate, collision_rate_sum, collision_end_time;
float select_prob, backoff_time, packet_slots;


int queue_size[MAX_STATIONS]; /* Current queue size at a station */
int int queue_id[MAX_STATIONS][MAX_Q_SIZE]; /* Array for packet ID's */
int id_list[ID_SIZE]; /* Array of id_numbers */
int id_attempt_stn[MAX_STATIONS]; /* Array for attempting stations */

int i, j, ic, ii, next_station, next_event, next, id_number;
int no_attempts, no_trans, no_collisions, select_flag;


main ()
{
printf("The following results are for:  \n");
printf("Degrees of freedom = %d\n", DEGREES_FR);
printf("Confidence Interval = 95 percent \n");
printf(" ========================================= \n");
printf("\n");

arrival_rate = 0.0;
slot_size = BUS_LENGTH * FACTOR * 5.0 * pow (10.0, -9.0);
p = PERSIST;
packet_time = PACKET_LENGTH * FACTOR / BUS_RATE;
packet_slots = (float) (int) (packet_time/slot_size) + 1.0;
infinite = 1.0 * pow (10.0, 30.0);
rand_size = 0.5 * pow (2.0, 8.0 * sizeof(int));

for (ii=0; ii < 10; ii++)
         {
         arrival_rate = arrival_rate + 20.0;
         for (ic = 0; ic <= DEGREES_FR; ic++)
         {
rho                       = 0.0;
ch_busy                   = 0.0;
clock                     = 0.0;
d_clock                   = 0.0;
collision_end_time        = 0.0;
utilization               = 0.0;
no_pkts_departed          = 0.0;
total_delay               = 0.0;
next_event_time           = 0.0;
average_delay             = 0.0;
no_collisions             = 0;
select_flag               = 0;

/* Compute the traffic intensity.  If the traffic intensity
      is greater than unity, stop the program.  */

rho = arrival_rate * PACKET_LENGTH * MAX_STATIONS / BUS_RATE;

if (rho >= 1.0)
{
printf("Traffic intensity is too high");
exit(1);
}

/* Initialize all variables to their appropriate values.  */

arrival_rate_slots = arrival_rate * slot_size;
      for (i = 0; i < MAX_STATIONS; i++) queue_size[i]=0;
      for (i = 0; i < ID_SIZE; i++)
        {
        start_time[i] = 0.0;
        id_list[i] = 0;
        }
```

```
      for (i = 0; i < MAX_STATIONS; i++)
        {
        for(j = 0; j < MAX_Q_SIZE; j++) queue_id[i][j]=0;
      }

for (i = 0; i < MAX_STATIONS; i++)
        {
        for (j = 0; j < 4; j++)
         {
         event_time[i][j] = infinite;
         x = (float) rand();
         x = x * FACTOR/rand_size;
         if (j == 0) event_time[i][j] = x;
         }
        }

/* Scan the event list and pick the next event to be executed.  */
while (no_pkts_departed < MAX_PACKETS)
        {
        next_event_time = infinite;
        for (i = 0; i < MAX_STATIONS; i++)
         {
         for (j = 0; j < 4; j++)
          {
          if (next_event_time > event_time[i][j])
           {
           next_event_time = event_time[i][j];
           next_station = i;
           next_event = j;
           }
          }
         }
clock = next_event_time;
 if (next_event > 3)
        {
        printf("Check the event list");
        exit(1);
        }
 while (d_clock <= clock) d_clock ++ ;
 switch (next_event)
 {
case 0:  /* This is an arrival event.  */
    {

    /* Select an identification for the arriving message */
    id_number = -1;
    for (i = 0; i < ID_SIZE; i++)
      {
      if (id_list[i] == 0)
        {
        id_number = i;
        id_list[i] = 1;
        break;
        }
      if (id_number != -1) continue;
      }
    if (id_number == -1)
      {
      printf("Check the ID-list.");
      exit(1);
      }
```

```
    queue_size[next_station] ++ ;
    if (queue_size[next_station] > MAX_Q_SIZE)
      {
      printf("The queue size is large and is = %d\n",
      queue_size[next_station]);
      exit(1);
      }
    queue_id[next_station][(queue_size[next_station]-1)] =
    id_number;
    start_time[id_number] = clock;
    if (queue_size[next_station] == 1)
      {
      event_time[next_station][1] = d_clock;
      if (event_time[next_station][1] <= collision_end_time)
      event_time[next_station][1] = collision_end_time + 1.0;
      }

 /* Schedule the next arrival */

 for (;;)
         {
         x = (float) rand();
         if (x != 0.0) break;
         }
 logx = -log(x/rand_size) * FACTOR / arrival_rate_slots;
event_time[next_station][next_event] = clock + logx;
 break;
 }
 case 1:  /* This is an attempt event.  */
    {
    no_attempts = 0;
    for (i = 0; i < MAX_STATIONS; i++)
      {
      if (event_time[i][1] == clock)
        {
        no_attempts ++ ;
        id_attempt_stn[no_attempts - 1] = i;
        }
      }
    select_flag = 0;
    if (no_attempts > 1)
      {
      x = (float) rand();
      x = x/rand_size;
      for (i = 0; i < no_attempts; i++)
        {
        select_prob = (float) (i+1)/ ((float) no_attempts);
        if (x <= select_prob)
          {
          next_station = id_attempt_stn[i];
          select_flag = 1;
          }
        if (select_flag == 1) continue;
        }
      }
 if (ch_busy == 0.0)
    {
    if (p == 0.0)
      {
      event_time[next_station][2] = clock + 1.0;
      event_time[next_station][1] = infinite;
      }
```

```
    else
      {
      x = (float) rand();
      x = x/rand_size;
      if (x < p)
        {
        event_time[next_station][2] = clock + 1.0;
        event_time[next_station][1] = infinite;
        }
      else
        {
        event_time[next_station][1] = clock + 1.0;
        if (event_time[next_station][1] <= collision_end_time)
        event_time[next_station][1] = collision_end_time + 1.0;
        event_time[next_station][2] = infinite;
        }
      }
    }
if (ch_busy == 1.0)
    {
    if (p == 0.0)
      {
      x = (float) rand();
      x = x/rand_size;
      backoff_time = (float) (int) (x * MAX_BACKOFF);
      if (backoff_time < 1.0) backoff_time = 1.0;
      event_time[next_station][1] = clock + backoff_time;
      if (event_time[next_station][1] <= collision_end_time)
      event_time[next_station][1] = collision_end_time +
      backoff_time;
      event_time[next_station][2] = infinite;
      }
    else
      {
      event_time[next_station][1] = clock + 1.0;
      if (event_time[next_station][1] <= collision_end_time)
      event_time[next_station][1] = collision_end_time + 1.0;
      event_time[next_station][2] = infinite;
      }
    }
break;
}
case 2:  /* This is a transmission event.  */
    {
    no_trans = 0;
    for (i = 0; i < MAX_STATIONS; i++)
    if (event_time[i][2] == clock) no_trans ++ ;
    if (no_trans > 1)
      {
        {
        collision_end_time = clock + JAM_PERIOD + 2.0;
        no_collisions ++ ;
        }
      for (i = 0; i < MAX_STATIONS; i++)
        {
        if (event_time[i][2] == clock)
          {
          event_time[i][2] = infinite;
          x = (float) rand();
          x = x/rand_size;
          backoff_time = (float) (int) (x * MAX_BACKOFF);
          if (backoff_time < 1.0) backoff_time = 1.0;
          event_time[i][1] = collision_end_time + backoff_time;
          }
```

```
        if (event_time[i][1] <= collision_end_time)
          {
          x = (float) rand();
          x = x/rand_size;
          backoff_time = (float) (int) (x * MAX_BACKOFF);
          if (backoff_time < 1.0) backoff_time = 1.0;
          event_time[i][1] = collision_end_time + backoff_time;
          }
        }
      }
    else
      {
      if (ch_busy != 1.0)
        {
        event_time[next_station][3] = clock + packet_slots ;
        event_time[next_station][2] = infinite;
        ch_busy = 1.0;
        }
      else
        {
        if (p == 0.0)
          {
          x = (float) rand();
          x = x/rand_size;
          backoff_time = (float) (int) (x * MAX_BACKOFF);
          if (backoff_time < 1.0) backoff_time = 1.0;
          event_time[next_station][1] = clock + backoff_time;
          if (event_time[next_station][1] <= collision_end_time)
          event_time[next_station][1] = collision_end_time +
          backoff_time;
          event_time[next_station][2] = infinite;
          }
        else
          {
          event_time[next_station][1] = clock + 1.0;
          if (event_time[next_station][1] <= collision_end_time)
          event_time[next_station][1] = collision_end_time + 1.0;
          event_time[next_station][2] = infinite;
          }
        }
      }
    break;
    }
 case 3:  /* This is a departure event.  */
    {
    id_number = queue_id[next_station][0];
    ch_busy = 0.0;
    queue_size[next_station] -- ;

    /* Push the queue forward.  */

    for (i = 0; i < queue_size[next_station]; i++)
    queue_id[next_station][i] = queue_id[next_station][i+1];
    queue_id[next_station][queue_size[next_station]] = 0;
 delay = clock - start_time[id_number];
 total_delay += delay;
 id_list[id_number] = 0;
 no_pkts_departed += 1.0;
 utilization += packet_slots;
event_time[next_station][3] = infinite;
if (queue_size[next_station] > 0)
    {
    event_time[next_station][1] = clock + 1.0;
    if (event_time[next_station][1] <= collision_end_time)
    event_time[next_station][1] = collision_end_time + 1.0;
    }
```

```
else
    {
    event_time[next_station][1] = infinite;
    event_time[next_station][2] = infinite;
    }
    break;
    }
  }
}
      utilization = utilization / clock;
      average_delay = total_delay * slot_size / (no_pkts_departed * FACTOR);
      throughput = no_pkts_departed * FACTOR / (clock * slot_size);
      collision_rate = (float) no_collisions * FACTOR / (clock * slot_size);
      utilization_ci[ic] = utilization;
      delay_ci[ic] = average_delay;
      throughput_ci[ic] = throughput;
      collision_rate_ci[ic] = collision_rate;
      }
      delay_sum = 0.0;
      delay_sqr = 0.0;
      utilization_sum = 0.0;
      utilization_sqr = 0.0;
      throughput_sum = 0.0;
      collision_rate_sum = 0.0;
      for (ic = 0; ic <= DEGREES_FR; ic++)
            {
            delay_sum += delay_ci[ic];
            delay_sqr += pow (delay_ci[ic],2.0);
            utilization_sum += utilization_ci[ic];
            utilization_sqr += pow (utilization_ci[ic],2.0);
            throughput_sum += throughput_ci[ic];
            collision_rate_sum += collision_rate_ci[ic];
            }
      delay_sum = delay_sum / (DEGREES_FR + 1);
      delay_sqr = delay_sqr / (DEGREES_FR + 1);
      delay_var = delay_sqr - pow(delay_sum,2.0);
      delay_sdv = sqrt(delay_var);
      delay_con_int = delay_sdv * t_dist_par[DEGREES_FR-1]/sqrt(DEGREES_FR);
      utilization_sum = utilization_sum / (DEGREES_FR + 1);
      utilization_sqr = utilization_sqr / (DEGREES_FR + 1);
      utilization_var = utilization_sqr -pow(utilization_sum,2.0);
      utilization_sdv = sqrt(utilization_var);
      utilization_con_int = utilization_sdv *
            t_dist_par[DEGREES_FR-1]/sqrt(DEGREES_FR);
      throughput_sum =
      throughput_sum / (DEGREES_FR + 1);
      collision_rate_sum = collision_rate_sum / (DEGREES_FR + 1);
      printf("For an arrival rate = %g\n",arrival_rate);
      printf("The average delay = %g", delay_sum);
      printf(" +- %g\n", delay_con_int);
      printf("The utilization = %g", utilization_sum);
      printf(" +-%g\n", utilization_con_int);
      printf("The throughput = %g\n", throughput_sum);
      printf("The collision rate = %g\n", collision_rate_sum);
      printf("\n");
      }
}
```

Appendix C

Simulation Program for Star LANs

```
/* This program simulates star local area networks.  */

# include    <stdio.h>
# include    <stdlib.h>
# include    <math.h>


# define   MAX_STATIONS          50          /* Number of stations */
# define   RATE                  5000000.0   /* Transmission rate in
                                             bits per second */
# define   PACKET_LENGTH         1000.0      /* Packet length (bits) */
# define   POLLING_PKT_LENGTH    50.0        /* Poll packet length (bits) */
# define   BUS_LENGTH            1000.0      /* Bus length in meters */
# define   MAX_PACKETS           1000        /* Maximum packets to be
                                             transmitted in simulation run */
# define   FACTOR                1000.0      /* A factor used for
                                             changing units of time */
# define   MAX_Q_SIZE            100         /* Maximum queue size */
# define   DEGREES_FR            5           /* Degrees of freedom */

float arrival_rate; /* Arrival rate (in packets/sec) for each station */
float tau; /* End-to-end propagation delay */
float packet_time; /* Packet transmission time */
float polling_pkt_time; /* Polling packet transmission time */
float start_time [MAX_STATIONS][MAX_Q_SIZE]; /* Starting time of packets */
float event_time [MAX_STATIONS][3]; /* Time of occurrence of an event */
float t_dist_par [10] = {12.706, 4.303, 3.182, 2.776, 2.571, 2.447, 2.365,
                 2.306, 2.262, 2.228}; /* T- distribution parameters */
float delay_ci [DEGREES_FR + 1]; /* An array to store delay values */

float rho, clock, no_pkts_departed, next_event_time;
float x, logx, rand_size, infinite;
float delay, total_delay, average_delay;
float delay_sum, delay_sqr, delay_var, delay_sdv, delay_con_int;

int queue_size [MAX_STATIONS]; /* Current queue size at a station */
int i, j, ic, ii, next, next_station, next_event;
```

```
main ()
{
printf("The following results are for:  \n");
printf("Degrees of freedom = %d\n", DEGREES_FR);
printf("Confidence interval = 95 percent \n");
printf(" ========================================= \n");
printf("\n");

arrival_rate = 0.0;
tau = BUS_LENGTH * FACTOR * 5.0 * pow (10.0, -9.0);
packet_time = PACKET_LENGTH * FACTOR / RATE;
polling_pkt_time = POLLING_PKT_LENGTH * FACTOR / RATE;

for (ii=0; ii < 10; ii++)
        {
        arrival_rate += 20.0;
        for (ic = 0; ic <= DEGREES_FR; ic++)
         {

rho                   = 0.0;
clock                 = 0.0;
no_pkts_departed      = 0.0;
total_delay           = 0.0;
next_event_time       = 0.0;
average_delay         = 0.0;
infinite              = 1.0 * pow (10.0, 30.0);
rand_size             = 0.5 * pow (2.0, 8.0 * (float) size of (int));

/* Compute the traffic intensity.  If the traffic intensity
      is greater than unity, stop the program.  */

rho = arrival_rate * MAX_STATIONS / RATE;
if (rho >= 1.0)
        {
        printf("Traffic intensity is too high");
        exit(1);
        }

/* Initialize all variables to their appropriate values.  */

for (i = 0; i < MAX_STATIONS; i++) queue_size[i]=0;
for (i = 0; i < MAX_STATIONS; i++)
        {
        for (j = 0; j < MAX_Q_SIZE; j++) start_time[i][j]=0.0;
        }
for (i = 0; i < MAX_STATIONS; i++)
        {
        for (j = 0; j < 3; j++)
         {
         event_time[i][j] = 0.0;
         if (j !=0) event_time[i][j] = infinite;
         if (i==0 && j==2) event_time[i][j] = 0.0;
         }
        }

/* Scan the event list and pick the next event to be executed.  */

while (no_pkts_departed < MAX_PACKETS)
        {
        next_event_time = infinite;
        for (i = 0; i < MAX_STATIONS; i++)
         {
         for (j = 0; j < 3; j++)
          {
          if (next_event_time > event_time[i][j])
           {
```

```
            next_event_time = event_time[i][j];
            next_station = i;
            next_event = j;
            }
          }
        }
        clock = next_event_time;
if (next_event > 2)
        {
        printf("Check the event list");
        exit(1);
        }
switch (next_event)
 {
 case 0:  /* This is an arrival event.  */
        {
        queue_size[next_station] ++ ;
        if (queue_size[next_station] > MAX_Q_SIZE)
         {
         printf("The queue size is large and is = %d\n",
         queue_size[next_station]);
         exit(1);
         }
        start_time [next_station][(queue_size[next_station]-1)] = clock;

        /* Schedule the next arrival.  */

        for (;;)
         {
         x = (float) rand();
         if(x = 0.0) break;
         }
        logx = -log(x/rand_size) * FACTOR / arrival_rate;
        event_time[next_station][next_event] = clock + logx;
        break;
        }

        case 1:  /* This is a departure event.  */
         {
         queue_size[next_station] -- ;
         no_pkts_departed ++ ;
         delay = clock - start_time[next_station][0];
         total_delay += delay;

/* Push the queue forward.  */

if (queue_size[next_station] > 0)
        {
        for (i=0; i < queue_size[next_station]; i++)
         start_time[next_station][i] = start_time[next_station][i+1];
         }
start_time[next_station][queue_size[next_station]]=0.0;
event_time[next_station][next_event] = infinite;
break;
}

case 2:  /* Find the next station to be serviced.  */
         {
         event_time[next_station][next_event] = infinite;
         next = next_station + 1;
         if(next == MAX_STATIONS) next = 0;
         if (queue_size[next_station] > 0)
          {
          event_time[next_station][1] = clock + packet_time + tau;
          event_time[next][2] = clock + packet_time + polling_pkt_time
          + tau;
          }
```

```
        else
         {
         event_time[next_station][1] = infinite;
         event_time[next][2] = clock + polling_pkt_time + tau;
         }
        break;
         }
        }
}

average_delay = total_delay / (no_pkts_departed * FACTOR);
delay_ci[ic] = average_delay;
}
delay_sum = 0.0;
delay_sqr = 0.0;
for (ic = 0; ic <= DEGREES_FR; ic++)
        {
        delay_sum += delay_ci[ic];
        delay_sqr += pow(delay_ci[ic],2.0);
        }
        delay_sum = delay_sum / (DEGREES_FR + 1);
        delay_sqr = delay_sqr / (DEGREES_FR + 1);
        delay_var = delay_sqr - pow(delay_sum,2.0);
        delay_sdv = sqrt(delay_var);
        delay_con_int = delay_sdv * t_dist_par[DEGREES_FR-1] /
        sqrt(DEGREES_FR);
        printf("For an arrival rate = %g\n",arrival_rate);
        printf("The average delay = %g", delay_sum);
        printf(" +- %g\n", delay_con_int);
        printf("\n");
        }
 }
```

INDEX

T

U

V

W